Technocities

Technocities

edited by

John Downey and Jim McGuigan

HT
119
.T43x
1999
West

SAGE Publications
London • Thousand Oaks • New Delhi

Editorial arrangement © John Downey and Jim McGuigan
1999
Introduction © Jim McGuigan 1999
Chapter 1 © Stephen Graham 1999
Chapter 2 © Kevin Robins 1999
Chapter 3 © Frank Webster 1999
Chapter 4 © Simone Bergman and Liesbet van Zoonen 1999
Chapter 5 © Julian Stallabrass 1999
Chapter 6 © John Downey 1999
Chapter 7 © Leen d'Haenens 1999
Chapter 8 © Simon Bell 1999
Chapter 9 © John Pickering 1999
Chapter 10 © Douglas Kellner 1999
Afterword © John Downey 1999

First published 1999

All rights reserved. No part of this publication may be
reproduced, stored in a retrieval system, transmitted or
utilized in any form or by any means, electronic, mechanical,
photocopying, recording or otherwise, without permission in
writing from the Publishers.

SAGE Publications Ltd
6 Bonhill Street
London EC2A 4PU

SAGE Publications Inc
2455 Teller Road
Thousand Oaks, California 91320

SAGE Publications India Pvt Ltd
32, M-Block Market
Greater Kailash – I
New Delhi 110 048

British Library Cataloguing in Publication data

A catalogue record for this book is available from
the British Library

ISBN 0 7619 5555 0
ISBN 0 7619 5556 9 (pbk)

Library of Congress catalog card number 98-61655

Typeset by Photoprint, Torquay, Devon
Printed in Great Britain by Biddles Ltd, Guildford, Surrey

Contents

List of Contributors		vii
Acknowledgements		ix

Introduction 1
Jim McGuigan

Part 1 Debates

1 Towards Urban Cyberspace Planning: Grounding the Global through Urban Telematics Policy and Planning 9
   Stephen Graham

2 Foreclosing on the City? The Bad Idea of Virtual Urbanism 34
   Kevin Robins

3 Information and Communications Technologies: Luddism Revisited 60
   Frank Webster

Part 2 Textures

4 Fishing with False Teeth: Women, Gender and the Internet 90
   Simone Bergman and Liesbet van Zoonen

5 The Ideal City and the Virtual Hive: Modernism and Emergent Order in Computer Culture 108
   Julian Stallabrass

Part 3 Territories

6 XS 4 All? 'Information Society' Policy and Practice in the European Union 121
   John Downey

7 Beyond Infrastructure: Europe, the USA and Canada on the Information Highway 139
   Leen d'Haenens

8 Technocities and Development: Images of Inferno and Utopia 153
   Simon Bell

Part 4 Perspectives

9 Designs on the City: Urban Experience in the Age of Electronic Reproduction 168
   John Pickering

10 New Technologies: Technocities and the Prospects for
    Democratization   186
    Douglas Kellner

    Afterword: Back to the Future?   205
    John Downey

    Index   210

List of Contributors

Simon Bell lectures in the Systems Discipline in the Centre for Complexity and Change of the Open University, UK. He has worked for many years as a consultant on information and communication technologies projects in newly industrializing states. He is the author of *Sustainability Indicators: Measuring the Immeasurable* (1998).

Simone Bergman is a researcher in the Communications Department of Amsterdam University, the Netherlands.

Leen d'Haenens is Associate Professor at the Department of Communication, University of Nijmegen, the Netherlands, where she teaches European Media Policy, International Communication, and Media, Minorities and Prejudice. She recently co-edited *Media Dynamics and Regulatory Concerns in the Digital Age*, on Europe's media policy (1998). She also edited *Cyberidentities: Canadian and European Presence in Cyberspace*, to be published in 1999.

John Downey is Senior Lecturer in Communication, Culture, and Media at Coventry University, UK. He has published a number of articles on media and culture in Europe and is working on a book concerning the development of information and communication technologies in Europe. He also works as a consultant.

Stephen Graham is a Reader at the Centre for Urban Technology at the University of Newcastle, UK. He is the co-author of *Telecommunications and the City: Electronic Spaces, Urban Places* (1996, with Simon Marvin), and has written extensively about many aspects of city-technology relations.

Douglas Kellner is the George Kneller Chair in the Philosophy of Education at UCLA, USA, and is the author of many books, including *Critical Theory, Marxism and Modernity* (1989), *Jean Baudrillard* (1989), *Postmodern Theory* (with Steve Best, 1991), *Television and the Crisis of Democracy* (1990), *The Persian Gulf TV War* (1992), *Media Culture* (1995) and *The Postmodern Turn* (with Steve Best, 1997).

Jim McGuigan is Senior Lecturer in Sociology at Coventry University, UK, and is the author of *Cultural Populism* (1992) and *Culture and the Public Sphere* (1996). He has edited *Cultural Methodologies* (1997) and a thematic issue of the *International Journal of Cultural Policy* and is co-editor of *Studying Culture* (1993, 1997). His next book is *Modernity and Postmodern Culture*.

John Pickering lectures in the Psychology Department at Warwick University, UK. He has written widely on psychology, architecture and technology. His most recent publication is an edited collection, *The Authority of Experience* (1997).

Kevin Robins is Professor of Cultural Geography at the University of Newcastle, UK, and is the author of a number of books, including *Information Technology – a Luddite Analysis* with Frank Webster (1986), *Spaces of Identity: Global Media, Electronic Landscapes, and Cultural Boundaries*, with David Morley (1995), and *Into the Image* (1996).

Julian Stallabrass is Tutor in Art History and Theory at the Ruskin School, Oxford University, UK. He is the author of *Gargantua: Manufactured Mass Culture* (1996) and has written extensively about the aesthetics and ideology of information and communication technologies.

Frank Webster is Professor of Sociology at Birmingham University, UK. He is the author of several books, including *Information Technology – a Luddite Analysis* with Kevin Robins (1986), *Theories of the Information Society* (1995) and *The Postmodern University? Contested Visions of Higher Education in Society* (1997) with Anthony Smith. He recently completed *Times of the Technoculture* with Kevin Robins, and *Theory and Society: Understanding the Present* with Gary Browning and Abigail Halcli.

Liesbet van Zoonen is Senior Lecturer in Media Studies at the University of Amsterdam, the Netherlands. She is the author of *Feminist Media Studies* (1994) and recently edited *The Media in Question* with Kees Brants and Joke Hermes (1998).

Acknowledgements

This collection grew out of a conference organized by the Communication, Culture and Media subject group at Coventry University in March 1996. Much of the success of the conference may be attributed to the organizational work of Mike Walford who unfortunately could not be at the conference because of a family bereavement. Pierre Sacré stepped into the breach with aplomb, both before and during the conference.

We would like to thank the contributors to this volume, some of whom were not able to attend the conference but later volunteered to write a chapter. Publishing a book based on contributions which begin their lives at a conference is a difficult business and we acknowledge the value of the suggestions made by readers, most particularly Bob Catterall for his comments on the manuscript.

John Downey and Jim McGuigan

Introduction

Jim McGuigan

In the opening chapter to this book Stephen Graham provides a wide-ranging discussion of urban planning issues in relation to computer-mediated communications and cyberspace which usefully frames the general theme of the book as a whole. He takes issue with technological determinist perspectives whether of a utopian or dystopian kind and seeks, instead, 'to insert the idea of local agency into debates about new technologies and the future of cities' (p. 10). Like Raymond Williams (1974), Graham sees technological innovation and implementation as resulting from a multiplicity of factors that combine to produce specific effects in a variety of circumstances. There is no inevitability in the making and deployment of technologies and there is no place quite like any other place (see Lee, 1997). Global trends have different manifestations in different places. What happens always depends upon the particular conditions prevailing in context and the forces of business and political decision-making: hence, the stress on local agency. Many city governments around the world have seen great opportunities in information and communication technologies (ICTs) for urban regeneration and economic growth. Graham identifies a number of experiments that are concerned with 'global positioning', 'reconnecting the fragments' and, going beyond economistic reasoning, the communicational rights of citizenship.

The possibilities are illustrated by community cable networks, freenets, city host computers and virtual cities constructed on the World Wide Web. Amsterdam's *Digitale Stad* is an outstanding instance of how the city is reimagined in cyberspace and in a manner which relates to life in an actual city. This example is also discussed in Chapter 4, by Simone Berger and Liesbet van Zoonen, with regard to women's involvement. Graham himself observes that social disadvantage may not only be ameliorated by on-line services but can also be exacerbated by them, a theme which recurs throughout this book. For the physically immobile, for instance disabled and elderly people, cyberspace offers the prospect of improved control over their lives, yet material and cultural barriers are not necessarily easily overcome. And they are unlikely to be overcome where social campaigns and progressive public policies for extending access are absent (for a fuller discussion, see Graham and Marvin, 1996).

Running through this book is a tension between optimistic and pessimistic scenarios for the 'technocity' but, as Kevin Robins suggests in Chapter 2, the real tension may be rather more between those who see order – and hope for it – and those who see disorder. He is critical of the position outlined by Stephen Graham, whom Robins recognizes is not by any means the most starry-eyed of would-be optimists. The problem is, according to Robins, that the world imagined by urban planners, which is in turn inspired by the communicative resolutions of fragmentation and difference that are promised by the advocates of computer culture, is unconvincing. Behind this advocacy stands a corporate agenda of globalization which recycles old Enlightenment themes in a distinctly ideological fashion. Robins expresses 'incredulity and astonishment' at the techno-narratives that are being so widely circulated. In effect, then, the technocity may be seen as a distraction from the real conditions of cities throughout the world, overcrowded by enforced migration, where conflict and antagonism are constitutive. More positively, he sees these as disorderly places that have considerable appeal in comparison with the sanitized order of computerized rationality. There is a curious 'denial and disavowal' going on in the debate about informational cities, according to Robins, a failure to confront the terrors and the pleasures of actual cities. Robins has little doubt where the distracting visions are being imagined in the first place and what business and political interests they serve.

In the 1980s, Robins published a book with Frank Webster which criticized the information society thesis and was subtitled provocatively, *A Luddite Analysis* (Webster and Robins, 1988). Ten years on, Webster returns here to reconsider that critique and finds that there is not a great deal to reconsider. Technological utopianism has, in fact, become even more pronounced in the 1990s than it was in the preceding decade, and with no greater justification. During the intervening period, there have been theoretical developments and political developments that made the belief in an information technology fix increasingly potent. It has linked up with some of the more extravagant claims of postmodernist thought and been placed at the cutting edge of policy by both Bill Clinton's Democratic presidency and Tony Blair's New Labour government. Postmodernists have seen the Internet as the ideal medium of decentred subjectivity and the site where an endless play with one's identity becomes most possible (for instance, Poster, 1995). The convergence of this kind of thinking and conservative social theorizing such as that of Daniel Bell is striking, although, as Webster points out, it is the shallower writing of Alvin Toffler which has had the most popular impact or, at least, has articulated this strand of thought for politicians and 'policy wonks'. Communitarians and virtual communitarians have also contributed immensely to the intellectual 'excitement' associated with ICTs. In critiquing these ideas, Webster stresses 'the gulf between real social trends and the wishful thinking of many commentators' (p. 85).

The lines of influence and the ideological formation of a revamped 'information society' are traced by Webster in Chapter 3, particularly the way in which New Labour signed up to the agenda of the Clinton administration and kept within the bounds of the market reasoning laid down by the New Right regimes that the Democratic presidency in the USA and the Labour government in Britain succeeded. 'Globalization', with its international finance markets and 'flexible' labour markets displacing the powers and responsibilities of national government, facilitated by fast communications, frames the 'new' thinking, according to Webster, and he is devastatingly critical of how it masks the unreconstructed exploitative and divisive propensities of capitalism.

On a more optimistic note, however, there are city experiments from which a less compromised political message may be derived, such as in Amsterdam's Digital Stad. According to Simone Bergman and Liesbet van Zoonen, it is the closest thing to a technocity in existence. In their chapter, the main concern is not virtual urbanity in general as a technical achievement but the social demography of users, particularly female users of the Internet in general. While recognizing the enduring masculine dominance of ICTs, Bergman and van Zoonen resist a feminist pessimism concerning gender power and newer communications media. There is a need to study the 'presence' and not only the 'absence' of women. It has been asked: 'Does the Internet work differently if you are a woman?' (p. 97). Through preliminary interview research with female users Bergman and van Zoonen come to the conclusion that the Internet can function as 'the virtual translation of more or less traditional concerns of personal contact' (p. 105). Also, specifically, there is developing 'a lively "feminine" culture on the Internet' (p. 105). Bergman and van Zoonen find that these uses of ICT are most characteristic of comparatively young, well-educated and independent women. Their study is not a survey observing the criteria of statistical representativeness, but an exercise in interpreting what may be the increasingly typical feminine meanings of the Internet without going along, however, with the essentialist position that the Internet is somehow inherently feminine (Plant, 1997). Instead, the urgent task still, according to Bergman and van Zoonen, is to 'demasculinize' the Internet.

Karl Marx made a famous comparison between the bee and the architect: the busy bee is just a drone programmed to work; ideally, the architect is a creative agent conjuring up in his or her mind what is to be built in reality. Julian Stallabrass's metaphor for the construction of virtual cities in contemporary discourse and practice is the hive. It is curious that in a period which is supposed to be 'postmodern' the most successful business in the world should apparently exemplify modernist rationality. As Stallabrass remarks, 'Windows, the triumph of form over economy, poses as a rational system' (p. 109). The very iconography of cyberspace recalls the rational dreams of modernity, most notably the perfect cities envisaged by Le Corbusier, designed for order and speed. Surfing the net is a virtual reproduction of Le Corbusier's motorist speeding through the Radiant City of his imagining. Stallabrass also looks at computer art and particularly the work of its most celebrated British exponent, William Latham. Latham, in Stallabrass's words, 'constructs virtual objects from various pre-defined shapes, usually horns and tusks' that are modified by a program which simulates evolutionary processes (p. 112). It's only art but an art which mimics a modern fantasy of manipulation and has little, if anything, to do with a critical modernism that challenged the powers that be. Stallabrass asks somewhat disingenuously, from his point of view, but nonetheless incisively, '[i]s contemporary computer culture a haven of politically radical activity, anti-sexism, anti-racism and solidarity with Third World peoples?' (p. 115); and comes to the conclusion: not really. Much of computer culture in its leisure forms, particularly sexual play, is 'both deeply inconsequential and conservative'. The driving force of computer culture has more to do with capital accumulation than with communication, argues Stallabrass forcefully.

John Downey, in a rather different register, offers a critique of the dominant mode of thinking and policy-making concerning ICTs in the European Union which was signalled by the Bangemann Report for the European Council in 1995. The theme of 'catching up' with the USA and Japan is persistently reiterated and its accomplishment is thought about largely in free market terms. As Downey notes, 'what really seems to be driving policy is the goal of reducing leased-line costs to private industry thereby increasing their global competitiveness' (p. 126). From this perspective, the regulatory role of the nation state is diminished and those harbouring social democratic aspirations will find that the dominant position promises 'no more public money, subsidies or protectionism'. The situation, however, is somewhat more contradictory than the domain assumptions of neo-liberal hegemony and neo-technological determinism in policy discourse might suggest. Since Bangemann the EU has stressed 'inclusion', requiring national and local governments to take seriously their social responsibilities for the regulation and provision of ICT services.

Downey traces how academic debate has intersected with policy debate and sees particular value in the work of Manuel Castells (1996, 1997, 1998). Castells's membership of the High Level Group of Experts, which in its 1996 interim report criticized Bangemann, demonstrates that the debate is by no means closed. As Downey observes, '[a]lthough Castells accepts the vocabulary of the rupture, of a dramatic, revolutionary change sweeping through advanced capitalism as a result of developments in ICTs, he comes to significantly different conclusions from the techno-boosters who promise that there will be no losers' (p. 129). There is indeed an 'informational mode of production' developing in the single time-frame of a global economy which tends to separate the interests of capital from close alignment with those of nation states. There are winners and there are losers. In Western Europe, 'core' countries are comparatively rich informationally and technologically whereas the figures for Internet access and personal computers per head of population in 'cohesion' or 'peripheral' regions demonstrate huge disparities from the core. Castells has focused on the multiple forms of inequality associated with ICT deployment in Europe, the polarization of workforces between the professional-managerial beneficiaries and the socially excluded in 'dual cities' at the core itself. Key elements of progressive policy to ameliorate informational and attendant social inequalities, identified by Downey in his chapter, are the reconciliation of private interests with the rights of citizenship, trans-European resourcing and training programmes. A properly functioning public sphere in democratic societies is dependent upon equality of access and participation. In exploring the possibilities, Downey considers the case of Bologna as a city which has made strenuous efforts to achieve social inclusion in the use of ICTs.

When Al Gore mapped out the US government's perspective on the information (super)highway in 1994 the whole world paid attention. He stressed private investment, competition and flexible regulation tempered by open access and universality of service. These latter considerations are vital aspects of political decision-making by national and local governments in response to the cultural properties of communication technologies that transcend national borders and national identities. Questions of policy can up to a point be distinguished along hardware/software lines, between infrastructural development, on the one hand, and, on the other hand, specifically *cultural* policies (see McGuigan, 1996). Leen d'Haenens explores this distinction with regard to Europe, the USA, Canada and, to a lesser extent, Japan. There are complex questions to do with employment, intellectual property rights, cross-media ownership and antitrust legislation, privacy, censorship, security of electronic information and universal access. D'Haenens is primarily interested in the cultural aspects of policy; and, for her, Canada is an especially pertinent case due to its close proximity to the USA. It has long been a concern of Canadian governments to maintain a distinctive national identity whilst also negotiating the coexistence of English- and French-speaking populations.

A great deal of debate about cyberspace and technocities takes place in what used to be called 'the developed world' and most prominently in the USA. In recent years there has been a panic in Europe about keeping up with the new communication technologies (Bangemann, 1995). 'Tiger economies' in South East Asia are also on board the technology bandwagon. The meanings and implications of all this in 'the developing world' pose open questions. No nomenclature here is satisfactory, though 'North' and 'South' are now favoured terms which do not, however, correspond exactly to physical geography. Some countries in the South are moving forward while others are being left behind, especially in Africa; and in the North as well, particularly when post-communist Eastern Europe is taken into account, development is uneven.

Simon Bell writes here of his own experience of research in developing countries, Nigeria, Pakistan and China, and reviews the issues of technology transfer and technology as a spur to development. The question of modernization remains a matter of dispute and the problematical relations between the country and the city in poorer lands are still relatively little understood. In development studies the city has tended to be neglected in spite of continuing large-scale migration from the country to the city. The city in the South is a complex place where urban elites appropriate technologies and participate in 'modernization' while poor majorities become yet more marginalized. Again, there is a bifurcation of perspectives, fears envisaged in the city as 'hell' and hopes invested in the 'magnet' city. Bell's chapter aims to illuminate these tensions.

Marshall McLuhan once observed:

> In a culture like ours, long accustomed to splitting and dividing all things as a means of control, it is sometimes a bit of a shock to be reminded that, in operational and practical fact, the medium is the message. This is merely to say that the personal and social consequences of any medium – that is, of any extension of ourselves – result from the new scale that is introduced into our affairs by each extension of ourselves, or by any new technology. (1964: 15)

McLuhan went on to remark that, '[m]any people would be disposed to say that it was not the machine, but what one did with the machine, that was its meaning and message'. In his opinion, they were mistaken. For McLuhan, the technological medium really was determinate, bringing about transformations in the human sensorium. Most of the contributors to this book would tend broadly towards the opposite view due to an awareness of the power relations, structures and agencies that shape the development and use of technologies. Yet none would necessarily disagree with the argument that the properties of the medium itself have a measure of determinacy. John Pickering, in Chapter 9, explores this side of the debate in a long historical perspective and from a psychological point of view.

Biological evolution is a lengthy and drawn-out process: cultural evolution is much quicker. Take urban living, for instance: in 1900 only 10 per cent of the world's population lived in towns and cities; now, over half the world's population are urban dwellers. Pickering looks at how both urbanity and digital technology shape dramatic changes in psychological dispositions and everyday life. He draws upon Benjamin's unfinished arcades project (Buck-Morss, 1989) to consider these changes. Where the actual city had already created new kinds of sensibility, as Simmel insisted, the advent of virtual cities potentially brings about yet greater changes in the self and associative activity. Unfortunately, though, when studied from, say, the technologically utopian perspective of the MIT Media Lab, 'fundamental issues' are typically omitted, as Pickering argues, 'issues of empowerment, inclusion/exclusion and access' (p. 178). Pickering reviews many of the exciting applications of 'intelligent' machines, such as computer-aided design and architecture, but he ends on a rather pessimistic note, influenced by Baudrillard's (1983) theorizing of simulations and simulacra, which can be taken as a warning, perhaps against Baudrillard's own intent, about displacement of 'real world' issues concerning social involvement and participation.

In a concluding *tour de force*, Douglas Kellner aims to put it all in perspective or, rather, a multiperspectival framework. There is a need for a critical social theory that can account dialectically for the contradictory forces at play in what Kellner himself has named 'technocapitalism'. It is undoubtedly the case that capitalism integrates structures and practices, and functions through the newer technologies of information and communication. It is vital to perceive this coalescence in order to avoid either technological determinism or economic determinism; in this sense, Kellner's argument is reminiscent of Williams's (1974) critique not only of technological determinism but also, and frequently missed, of 'symptomatic technology', the argument that the technology is merely a symptom of something else, most typically capitalist and military machinations. Williams wanted to insert the mediating term of 'intention', that actual decision-making and struggle around what should be done matter; and, this is the space of politics. In effect, it is an argument for agency, which is the guiding principle of Stephen Graham's introductory argument. This is not, however, quite how Kellner sees it. He is concerned with the ideological and material determinations that set limits upon and open up possibilities for agency.

Kellner remarks: 'whenever there are new technologies people project all sorts of fantasies, fears, hopes, and dreams on to them, and I believe this is now happening with computers and new multimedia technologies' (p. 187). Kellner wants to find a third way between 'technophilia' and 'technophobia' but not an easy compromise or cosy consensus. He is extremely critical of the 'computopia' evinced by serious theorists as well as by politicians and propagandists. Yet, a deconstruction of the interests behind and the ideologies articulating the panacea of ICT is, for him, radically insufficient. He points to some interesting paradoxes, for instance the way in which, in the USA, cyberspace functions, to a degree, as a comparatively decommodified and, indeed, messy space in a heavily commodified and ordered culture. For reasons of educational policy, in the main, the state provides a certain amount of free access to the Net and there are computers that are freely available to those in education: universities, of course, but also schools. Moreover, radical movements such as the Zapatistas in Mexico act politically on-line (see Castells, 1997). In the end, Kellner argues for a theory and a politics that are both–and, not either–or.

References

Bangemann, M. (1995) *Europe and the Global Information Society*. Strasbourg: European Council.
Baudrillard, J. (1983) *Simulations*. New York: Semiotext(e).
Buck-Morss, S. (1989) *The Dialectics of Seeing – Walter Benjamin and the Arcades Project*. Cambridge, MA: MIT Press.
Castells, M. (1996) *The Rise of the Network Society*. Malden, MA and Oxford: Basil Blackwell.
Castells, M. (1997) *The Power of Identity*. Malden, MA and Oxford: Basil Blackwell.
Castells, M. (1998) *End of Millennium*. Malden, MA and Oxford: Basil Blackwell.
Graham, S. and Marvin, S. (1996) *Telecommunications and the City – Electronic Spaces, Urban Places*. London and New York: Routledge.

Lee, M. (1997) 'Relocating location – cultural geography, the specificity of place and the city habitus', in J. McGuigan (ed.), *Cultural Methodologies*. London, New Delhi and Thousand Oaks, CA: Sage.

McGuigan, J. (1996) *Culture and the Public Sphere*. London and New York: Routledge.

McLuhan, M. (1964) *Understanding Media*. London: Routledge & Kegan Paul.

Plant, S. (1997) *Zeros and Ones – Digital Women and the New Technoculture*. London: Fourth Estate.

Poster, M. (1995) *The Second Media Age*. Cambridge: Polity Press.

Webster, F. and Robins, K. (1988) *Information Technology – A Luddite Analysis*. Norwood, NJ: Ablex.

Williams, R. (1974) *Television – Technology and Cultural Form*. London: Fontana.

PART 1

DEBATES

1 Towards Urban Cyberspace Planning: Grounding the Global through Urban Telematics Policy and Planning

Stephen Graham

> The history of communications is not a history of machines but a history of the way the new media help to reconfigure systems of power and networks of social relations. Communications technologies are certainly produced within particular centres of power and deployed with particular purposes in mind but, once in play, they often have unintended and contradictory consequences. They are, therefore, most usefully viewed not as technologies of control or of freedom, but as the site of continual struggles over interpretation and use. (Murdock, 1993: 536–7)

> Why should we care about this new kind of architectural and urban design issue [the 'urban' design of cyberspace]? It matters because the emerging civic structures and spatial arrangements of the digital era will profoundly affect our access to economic opportunities and public services, the character and content of public discourse, the forms of cultural activity, the enaction of power, and the experiences that give shape and texture to our daily routines. (Mitchell, 1995: 5)

Vague notions of 'city-ness' and urbanism hold an important place in the current media hype and debate surrounding cyberspace and the Internet (Featherstone and Burrows, 1995). Popular commentary on the growth of telecommunications-based social interaction, shopping and information retrieval is peppered with the use of urban/spatial metaphors for describing the electronic spaces which people increasingly 'enter' and interact 'within'. Beyond the most obvious spatial metaphors – cyberspace, electronic frontier, information superhighway, website – remarks about 'cybercities', 'virtual cities', 'virtual communities', 'virtual shopping malls', 'cybercafés', and 'cybervilles' are increasingly common.

The importance of the urban/spatial is also growing in technological debates within academia. General debates about cyberspace, telematics and the future of cities are currently proliferating within disciplines as diverse as architecture, cultural studies, communications studies, science and technology studies, urban sociology and geography. Attention is increasingly directed towards exploring how the economic, social and cultural aspects of cities interact with the proliferation of advanced telematics networks in all walks of urban life (see, for example, Architectural Design, 1995; Mitchell, 1995; Shields, 1996). Here, the common 1980s assumption that the new communicational capabilities of telematics would somehow 'dissolve' the city has waned. Rather, it is now clear that cyberspace is largely an urban phenomenon. It is developing out of the old cities, and is associated with new degrees of complexity within cities and urban systems, as urban areas across the world become combined into a single, globally interconnected, planetary metropolitan system (Graham and Marvin, 1996). Research here now centres on the degree to which city economies can be maintained in a world of on-line electronic flows; the ways in which place-based and 'virtual communities' interact; and the related interactions between urban cultures rooted in traditional public spaces and 'cybercultures' operating within the virtual spaces accessed from computers (see Mitchell, 1995 and Graham and Marvin, 1996, for reviews).

Despite the central importance of the 'urban' in cyberspace debates, issues of urban policy and planning have been virtually absent within both popular and academic debates. Questions of agency and local policy tend to be ignored in the simple recourse to either generalized, future-oriented debates, or to macro-level, binary models of societal transformation. In these, new technologies are seen to be somehow autonomously transforming society *en masse* into some new 'information age', 'information society' or 'cyberculture'. The implication is that local councils, policy-makers and planners are little more than irrelevant, even anachronistic, distractions in this exciting and epoch-making transformation.

With utopianism and crude technological determinism often dominating popular (and, in many cases, academic) debates, it is not surprising that the potential roles of urban policy-makers and planners in 'socially shaping' new technologies in cities at the local level are usually overlooked. This neglect, however, is problematic. It means that a fast-growing wave of urban experimentation with telematics, which is emerging across advanced industrial cities, is almost completely ignored. This is a problem, because such innovation promises to have major practical and theoretical implications for how we might consider the future of cities, urban policy and planning. It may also offer lessons on the broader question of how we might best understand the relations between cities and telematics, and how we might address the crucial question of thinking about the 'local' and the 'urban', in an increasingly telemediated and globalized era.

This chapter attempts to help insert the idea of local agency into debates about new technologies and the future of cities. It has three sections. In the first I try and explain why the concept of local agency has been so ignored in the rhetoric about cyberspace, cyberculture and the many allegations that we are moving toward a more telemediated society.

In the second section, I build on this discussion to explore the recent wave of experimentation at the urban level with information technology and telematics. Many of these initiatives are attempting to use telematics to help underpin the emergence of the more socially 'progressive', culturally enlivening and economically beneficial scenarios at the urban level. Three broad areas of such policies are discussed: 'global positioning' policies aimed at projecting a city as a global node for investment; internally focused telematics initiatives aiming at 'reconnecting the fragments' that increasingly characterize cities; and strategies aimed at developing electronic linkages between citizens and municipalities. I conclude, in the third section, by assessing the significance of these policies for our treatment of the 'local', for our understanding of cities, and for our conceptualization of telematics-based innovation more broadly.

Explaining the Neglect of Local Agency in Debates about Cyberspace and Cities

Why are general debates about cyberspace and the future of cities so buoyant, whilst the idea of there being local 'manoeuvring space' to shape local telematics development in cities is so rarely stressed? This, I argue, can be attributed to two problems. First, the dominant models of technology–society relationships which underpin cyberspace and city debates (technological determinism, futurism/utopianism and dystopianism/political economy) operate to deny the very concept that local agency can shape technological innovation in diverse and contingent ways. Secondly, the urban studies and policy communities themselves have been very slow to become aware of telematics.

Technological Determinism, Utopianism and the 'Candy Store Effect'

As with the wider discussion of technology–society relationships, the analysis of the linkages between cities and telecommunications tends to be dominated by a set of approaches which can broadly be termed 'technological determinism'. More often than not, in this 'mainstream' of social research on technology (Mansell, 1994), new telecommunications technologies are seen as direct causes of urban change (Edge, 1988: 1). This is because of their intrinsic qualities or 'logic' as space-transcending communications channels. The forces that stem from new telecommunications innovations are seen to have some autonomy from social and political processes (Winner, 1978).

Here, the social and the technical are cast as two different arenas, the former being shaped by the latter. Machines and technologies are seen to arise and evolve in a separate realm to alter the world (Thrift, 1993).

Technological 'revolutions', such as the current one which many allege to be based on telematics, are seen as virtually unstoppable broad waves of innovation and technological application, which then go on to 'impact' on cities and urban life (Miles and Robins, 1992). As with much social research on technology, literature on telecommunications and cities still tends to invoke what Gökalp (1992) calls 'grand metaphors' of the nature of telecommunications-based change in cities.

Invariably, modern telecommunications are seen as a 'shock', 'wave' or 'revolution' impacting or about to impact upon cities. In these scenarios, current or future urban changes are often assumed to be determined by technological changes in some simple, linear cause-and-effect manner. The use of simple two-stage models to describe changes in cities and society is common. Cities are placed in a new 'age' in which telecommunications increasingly have a prime role in reshaping their development. Most usual here are notions that capitalism is in the midst of a transformation towards some 'information society' (Lyon, 1988) or 'postindustrial society' (Bell, 1973), or that a more general 'communications revolution' (Williams, 1983) or 'third wave' (Toffler, 1980) is sweeping across urban society. The broad 'technological cause – urban impact' approach reflects very closely the 'commonsense' view of technological change within Western culture. As Stephen Hill argues:

> the experience of technology is the experience of apparent inevitability . . . the most influential critics who have sought to understand the experienced 'command' of technological change over twentieth-century life have turned to the machines for explanation, and asserted the 'autonomy' of technology. . . . The technological determinist stance aligns with many people's everyday experience. (Hill, 1988: 23–4)

Most often, though, because of the general inability to analyse real change and the influence of futurology, analysis centres on speculation concerning the 'impacts' of such telecommunications 'revolutions' on future cities in a general and vague way. As Kevin Robins suggests, 'all this is driven by a feverish belief in transcendence; a faith that, this time round, a new technology will finally and truly deliver us from the limitations and the frustrations of this imperfect world' (Robins, 1995: 136). The speculations of 'futurologists', and many 'cyber utopians', generally tend to take an optimistic view of the future 'impacts' of telecommunications on cities and urban life, offering tantalizing glimpses of future scenarios. The proliferation of electronic spaces and networks is often seen to be leading to an alternative reality which offers potential for 'recreating the world afresh' (Robins, 1995: 153).

In this rush to describe this re-creation of the world, actual telecommunications-based developments in real contemporary cities are rarely analysed in detail. If virtual spaces are mythologized as some 'point of departure' for society, then attention will always be deflected from the detail of how they relate to real people, real economies and real communities in real cities. A related tendency is to assume that the 'impacts' of telecommunications on cities are all the same. In fact, the immaturity and neglect of urban telecommunications studies means that there has been a tendency to approach the whole subject without trying to justify the theory or methodologies adopted. In the excitement to address these neglected and important areas, Warren (1989) notes what he calls a 'candy store effect':

> The topic [of telematics and urban development] creates a 'candy store' effect by providing license to deal with a range of phenomena. The result is an effort to cover far too much with no logic or theory offered to explain why some consequences are discussed and others are not and why some evidence is presented and other findings are not. . . . We are left with an analysis which lacks any theoretical base and an explicit methodology, gives more attention to marginal than primary effects of telematics, and, in many instances, is in conflict with a significant body of research. (Warren, 1989: 339)

The crucial point for this discussion is that this stress on autonomous technology, transcendence, positive scenarios and future cities suggests that analytical and policy debates centre around how society can adapt to and learn to live with the effects of telecommunications-based change, rather than focusing on the ways in which these effects may be altered or reshaped through policy initiatives. Implicitly, little space is left within these approaches for forces of human agency or urban and telecommunications policy-making at the local level with which to alter the apparent 'destiny' embodied in the telecommunications-based development of a city. As Robert Warren continues, 'benign projections give little indication that there are significant policy issues which should be on the public agenda' (Warren, 1989: 345). Kevin Robins and Mark Hepworth elaborate on this point, arguing that

> it is this question of agency that is fundamental. Within this futuristic scenario, technology appears to have its own autonomous and inevitable force. . . . It is a force, moreover, that becomes associated with a higher state of human evolution. . . . Insofar as technological development seems inflexible and unquestionable, and the course of progress to be part of quasi-evolutionary destiny, then perhaps the only appropriate response is that of acquiescence and compliance. (Robins and Hepworth, 1988: 157)

Dystopianism and Social Determinism

A second range of analytical approaches to city–telecommunications relations which often also implicitly deny the potential for urban telematics policies can be grouped under the broad heading of dystopianism and urban political economy. Dystopians and political economists stress the ways in which the development and application of telematics technologies are not somehow separate from society. Rather, they are seen to be fully inscribed into the political, economic and social relations of capitalism. Following this, telecommunications are not seen as simple determinants of urban change. Nor are they cast as panaceas or 'quick fix' technical solutions to urban problems. According to this approach, city–telecommunications relations cannot be understood without considering the broader political, economic, social and cultural relations of advanced industrial society and how they are changing.

Dystopians have been influenced by several strands of work. 'Postmodern' urban dystopias – from Ridley Scott's portrayal of Los Angeles in *Blade Runner* to William Gibson's 1984 *Neuromancer* – mix with more analytical strands of research on urban political economy (Castells, 1989), the political economy of telecommunications (Sussman and Lent, 1991), and analysis of the urban/technological changes involved in the shift to postmodernism (see, for example, Knox, 1993). These approaches offer disturbing anti-utopian or 'dystopian' visions and analysis of telecommunications-based urban life today and in the future (e.g. Davis, 1990; 1992; Virilio, 1993; Brook and Boal, 1995).

Political economy has a great deal more to commend it as a foundation for understanding telecommunications in cities than technological determinism or futurism (see Castells, 1989). This perspective centres on how society influences technology rather than the other way round; the 'effects' of telecommunications on cities are defined by the ways in which they are used to support wider processes of economic, political and spatial restructuring. Above all, especially in the neo-Marxist accounts, the development and application of telematics are seen to be driven by the imperative of maintaining capital accumulation for firms, and the need to overcome crises that reflect capitalism's inherent contradictions. City–telecommunications relations are, in this approach, seen to be driven largely by the economic forces surrounding the globalization of capitalism itself and to reflect and perpetuate capitalism's highly unequal social relations.

As with utopian accounts, however, critical theorists can easily become totalizing in predicting some bleak urban future/present. Ironically, they often end up agreeing with many utopians in their prediction of the inevitable end of the place-based meaning of the city. Michael Sorkin (1992: xi) believes that 'computers, credit cards, phones, faxes, and other instruments of instant artificial adjacency are rapidly eviscerating historic politics of propinquity, the very cement of the city'. Paul Virilio, meanwhile, predicts a total collapse of the physical, public aspects of cities. Urban residents, he argues, will soon become saturated with home-based, telemediated experiences accessed via interactive prostheses. All movement will be avoided as people fall victim to some all-encompassing 'domestic enslavement' – what he calls the life of the 'motorized handicapped' (Virilio, 1993: 11).

But many such approaches can be criticized for social rather than technological determinism. Often, the neo-Marxist accounts are economistic – their stress falls too heavily on the all-powerful influence of the globalizing political-economic structures of capitalism in determining telematics developments. Such social determinism, again, serves to reduce the apparent 'manoeuvring' space left for local innovation in telematics to alter the apparent destiny of the powerless urban locality stricken by social polarization and financial collapse. In this scenario, city planning and policy-making become little more than functional local agents which are coerced into providing the needs required to tempt multinational capital into an urban area. In the more reductionist accounts, cities are often portrayed as little more than 'helpless pawns of international corporate elites' (Judd and Parkinson, 1990: 14), which are coerced by the globalization of capital to do little but compete relentlessly for corporate and state investments, favourable imagery and tourist and conference visitors. David Harvey, for example, argues that

> we here approach a force that puts clear limitations upon the power of specific projects to transform the lot of particular cities. Indeed, to the degree that inter-urban competition becomes more potent, it will almost certainly operate as an 'external coercive power' over individual cities to bring them closer into line with the discipline and logic of capitalist development. (Harvey, 1989: 10)

The Neglect of Telematics in Planning and Urban Policy Debates

A third factor which helps to explain the low profile of local policy in telematics and city debates is the neglect of communications infrastructures within urban policy and studies communities themselves (Graham, 1992, 1994). Telecommunications remain perhaps the single most underdeveloped area of urban studies and urban planning and policy-making (Batty, 1990a). Communications studies, meanwhile, has long neglected the city as a focus of research (Jowett, 1993).

The effects of this divorce between urban studies and policy-making on the one hand, and communications studies and policy-making on the other, were compounded by the national dominance of telecommunications regulation, and their relative invisibility in cities. The result was that 'the concept of an urban communication infrastructure was not expressed in local institutions comparable to those which realised public concerns with education, environmental quality, housing and transportation. . . . Communication is not usually treated as part of the local technical infrastructure of urban life' (Mandelbaum, 1986: 132–5). Whilst this is changing quickly with the emergence of new research, and the explosion of highly visible telematics networks like the Internet and cable networks, urban policy-makers attempting to intervene in telematics are usually doing so for the first time, and with very little knowledge or experience.

There are problems here because the foundations of planning education and skills still show the legacy of planning's origins in focusing largely on the physical mobility of people and goods within cities, and the location of physical facilities and land uses within a unitary, integrated city (Webber, 1968: 1093). Electronic forms of communications are rarely discussed, the implication being that they are ubiquitous, invisible and of little importance in shaping the city. To Henry Bakis (1995: 3), telecommunications remain peripheral to urban and regional planning because of the 'persistence of the traditional paradigm whereby the approach to regional development remains, to a large extent, based on the logic of industrial development'.

But the growing importance of electronic interactions, information exchanges and transactions within and between cities has led some commentators to talk of a 'paradigm crisis' in urban planning and policy-making, because, they argue, the old ideas of planning the 'industrial city' are not appropriate to planning the 'information city' (see Graham and Marvin, 1996). Ken Corey has commented that 'urban and regional planning practice throughout many of the world's industrial market economies is in a state of paradigm challenge. In essence, the crisis exists because old planning procedures of how the industrial city functions don't seem to apply for today and tomorrow' (Corey, 1987: 121).

The result is that, even with a massive current growth in urban communications policies, urban policy-makers often still tend to remain wedded to crude versions of the 'grand metaphors' for explaining the shift to a more telecommunications-based society. Simple technological determinism is common. Most have, at best, a crude understanding of the technological and regulatory shifts that are under way in telecommunications. And because urban politicians and planners still remain firmly wedded to the tangible and salient aspects of cities, the arcane, mysterious and intangible world of telecommunications presents major problems as a focus of intervention. A senior politician within a UK city recently admitted that 'within the council, we've got a lot of bright local politicians, but they're not very good at working within the conceptual frameworks [of telecommunications]; they like to touch and feel, to know what's happening. They like to be very practical' (Graham, 1997: 245). There are some signs that this is slowly changing, however. The incorporation of advanced telecommunications into urban and regional planning has probably developed furthest in France. A recent communications plan for the French city of Lille, for example, comments that

> the traditional concepts of urban and regional planning are today outmoded. The harmonious development of areas towards equilibrium, the correct sharing out of resources, providing support to complementary developments within the city . . . these ideas have given way to the impression that spaces are fragmented, atomised and strongly competitive. . . . The insertion of telecommunications into the city makes the development of spaces more complex and introduces today a third dimension into urban and regional planning [after space and time]: this is the factor of real-time. (ADUML, 1991: 3)

Current Innovation in Urban Cyberspace Planning

The last decade has seen a worldwide upsurge in urban attempts to use telecommunications as policy tools for economic, social and cultural development. That this has occurred despite the powerful influence of futurology, technological determinism and dystopian scenarios, and the slow growth of debate on telecommunications in urban studies and policy-making is a measure of the magnitude of urban crises across the advanced industrial world. Telematics has become a natural policy focus as policy-makers everywhere have struggled to reinvigorate city economies, physically regenerate urban areas, market urban spaces as global sites for investment, address social polarization, and restructure public services to address funding crises (see Healey et al., 1995). Three areas where telematics and telecommunications have emerged as key policy focuses can be highlighted:

- The 'global positioning' approach, where telematics are used to attract inward investment into cities;
- The 'endogenous' development approach where telematics are used to try and 'reconnect' the economic, social and cultural fragments that increasingly characterize contemporary cities; and
- The delivery of public services via telematics and the establishment of new channels of city–citizen communication.

Global Positioning: Teleports and Urban Marketing Strategies

As the global shift towards market-based development of telecommunications gains ground, an increasing number of urban telematics strategies in the USA, the UK and Western Europe are emerging aimed at positioning cities as attractive global investment nodes for advanced service and manufacturing industries. As in the national-led 'future city' strategies such as those that developed in the 1980s in Singapore, France and Japan, gaining better telecommunications infrastructures than competing cities, or at least generating the *perception* of better infrastructure, is seen as an increasingly important policy objective for city authorities in these nations (Batty, 1990b).

The result of these shifts is rapid current growth in urban strategies aimed at using telecommunications to improve the economic competitiveness of individual cities as sites for the operation of global and multinational corporations. At the same time, there is a tendency for debates about 'urban regeneration' to centre on the need for property-led initiatives, and on the perceived social, cultural, economic and environmental needs of corporate business elites (Imrie and Thomas, 1993). The much-vaunted 'public–private' partnerships behind these initiatives often represent what Derek Shearer calls the 'edifice complex'. Here, he argues, ' "progress" is equated with 'the construction of high-rise office towers, sports stadia, convention centres, and cultural megapalaces, but ignores the basic needs of most city residents' (Shearer, 1989: 289).

Teleports – satellite links associated with property developments and links to local telecommunications networks – are one such body of initiatives. The 30 or so operational teleports in Western cities actually involve a variety of different kinds of initiative (IBEX, 1991). But most teleports effectively consist of nodes for advanced national and international telecommunications services, implanted into a part of a city and linked to local telecommunications networks for distributing access to the services locally. Often teleports form the centrepiece of ambitious urban redevelopment plans, with new office, industrial and high-status housing property developed around the facility. This is the case in Amsterdam, New York and Cologne.

Because of their catalytic function, and their potent marketing potential (Richardson et al., 1994), municipal authorities see teleports as potential centres of excellence and innovation in business telematics. Their 'high tech' and 'switched in' imagery is especially useful to certain cities suffering the effects of industrial decline. In Roubaix, France, for example – the town with the worst inner city crisis in the whole country – an ambitious teleport-based redevelopment plan goes under the banner of the phrase 'On the networks of the future' (Graham, 1995a).

Teleports aim to emerge as centres for the diffusion of innovation into the wider urban economy, through the creation of 'hot spots' of telematics demand in these key sectors, as well as providing new linkages between the urban and the global economy. This may improve economic competitiveness and the chances of attracting inward investment as a result of linking these new services to the needs of key sectors of the city economy. Often, teleports are linked with the construction of sophisticated optic fibre networks within cities, known as Metropolitan Area Networks (MANs), distributing access to high-bandwidth telematics services to large firms, higher education institutions and large government organizations. A common approach is to tailor the teleport to the specialized needs of a key local economic sector, whether it be media industries (Cologne Media Park), broadcasting (London Docklands teleport), financial services (New York and Edinburgh teleports), textiles (Roubaix teleport), high technology research and development (Sophia Antipolis in France) or maritime, port and logistics industries (Le Havre maritime city initiative and Bremen teleport).

A more recent set of global positioning strategies, which is bound up with the explosive growth of the World Wide Web, is the construction of web-sites aimed at marketing cities to elite tourists and conference organizers. A growing range of cities are setting up hosts and servers on the Internet containing maps, information, photographs, transport information, information on arts and culture, and guides to their facilities which real and potential visitors may access. Closely linked with the range of 'virtual cities' now under development, such web-sites make the most of the multimedia performance of the World Wide Web to market cities as products in the global 'image space' of the Web (Graham and Marvin, 1996). They aim to construct convivial, cosmopolitan and animated images for cities in the minds of influential decision-makers in conference bureaux, inward-investing firms and the highly mobile, largely elite tourist classes that dominate use of the World Wide Web. Over 2,000 cities are linked up into a specialist, global World Wide Web network called 'City.Net' through which Internet users can 'visit' a list of cities all over the world just by clicking their names. Not surprisingly, City.Net is geared to the international travelling elites who dominate use of the Internet.

Reconnecting the Fragments: Telematics for Endogenous Economic, Social and Cultural Development

There are some signs that the 1980s fashion for large, infrastructure-led telecommunications strategies such as teleports is being eclipsed by strategies aimed at using telematics as tools of endogenous urban planning and policy. Here the aim is to construct telematics applications which help 'reconnect' the economic, social, geographical and cultural fragments that increasingly characterize contemporary cities. The logic here is that telematics applications supporting computer-mediated communication, information exchange and transactions may help reinvigorate and reintegrate the 'local' when all trends seem to be fracturing metropolitan regions within a globalizing world.

This, the theory goes, may help underpin the development of a 'virtuous circle' where improved social cohesion is linked with a renaissance of urbanism, local economic development and civic culture. In this scenario, telematics may help to break down the barriers and fear between social, cultural and geographical groups within cities. They may help ground urban community development in an increasingly globalized context. They might support new telemediated dynamics of local economic growth within the city. And they might generate new spaces for interaction, debate and cultural development which feed back positively to help support a renaissance in urban social and cultural life.

Key here is the complex blending of place-based and placeless 'virtual communities'. Some argue that the multitude of specialized virtual communities on the Internet is evocative of a sense of convivial urbanism that has often been lost in the physical and social transformations towards postmodern urbanism (Rheingold, 1994). This is critical because social networks and the ties between people and places now regularly transcend the often arbitrary definitions of 'neighbourhoods' and 'the city'. In recognizing this urban fracturing, locally based telematics initiatives may help to 'attach' virtual communities to the dynamics of individual cities. Geoff Mulgan, for example, argues that 'given that the architecture and geography of large cities and suburbs has dissolved older ties of community, electronic networks may indeed become tools of conviviality within cities as well' (Mulgan, 1991: 69). He urges local policy-makers to explore this potential by adding telecommunications polices to their more familiar remit on education, planning, transport and housing.

This is increasingly happening. Four broad types of initiative are emerging from urban policy innovations to use telematics to boost the endogenous development of cities: community cable networks, videotex applications, 'freenets', and city host computers and 'virtual cities' based on the World Wide Web. On the cable front, a wide range of community access initiatives involving cable have recently been developed by city authorities in the USA. Using the First Amendment and local regulatory powers, cable access programming involving channels dedicated to community TV production is now fairly common. These channels can be considered as 'electronic public spaces', which offer a counter to the commercial imperatives of the marketplace (Aufderheide, 1992). These policies, however, have resulted from long battles against the centralization of control over cable and the overwhelming power of the market imperative in shaping their development. Only 10–15 per cent of US cable networks have such facilities. Far from being an automatic benefit of the supposed diversity of cable services, they have developed only where 'municipalities and citizens carve out public spaces with ingenuity, against the odds, and [are] rarely noticed in the national media' (Aufderheide, 1992: 53; see Strover, 1989).

Similar ideas are supporting a wide variety of videotex experiments in France linked with the ubiquitous Minitel system. The initiatives of the municipality of Marne la Vallée, near Paris, are typical. Aiming to structure a whole new set of 'modern public spaces' based on publicly accessible information and communication services on Minitel, the city authority has made efforts to develop a diversity of local social applications on the system. This is an attempt to widen participation in telematics as much as possible. The hope is to support new approaches to local democracy, through Minitel-based networks between citizens, centres of education, social organizations and municipal departments (Weckerle, 1991).

Electronic 'freenets' use commercial and municipal sponsorship to develop freely accessible electronic civic telematics networks. These services, which offer electronic mail, conferencing, information services, bulletin boards, and – often – wider Internet access, are becoming increasingly common in North American cities (Winner, 1993). There were over 25 at the end of 1994; eight others were being developed in Western Europe. Special equipment is often provided for people with various forms of disabilities to use freenet services. The objective of 'reconnecting' the fragments of cities is shown by the fact that many freenets are actually set up with structures that are analogous to the different physical elements of cities themselves. Complementing American freenets is a wide range of simple urban bulletin board services (or BBS), most delivered by computer enthusiasts from their own personal computers. It is estimated that there are 300 BBS services in Los Angeles alone. SF Net, a BBS service with 3,000 regular users in San Francisco, has set up 20 coin-operated terminals in cafés across the city. These are aimed at people without normal access to personal computers and modems.

Finally, a wide range of other 'electronic public spaces' are being developed, based on networks of municipally controlled host computers and 'virtual city' websites (Graham, 1995b). The Manchester Host, for example, offers a wide range of electronic mail, bulletin boards and database services to registered users through computers attached to phone lines. Whilst its services are charged for – unlike those of freenet – excluded groups are being supported through a network of 'electronic village halls' in the city. These are physical centres where training and host services are supported for 'communities of interest' – for example the Bangladeshi community, women's organizations, disabled groups and old people – and also distinct geographical communities.

But the most influential, and fastest growing initiatives, currently, are the so-called 'virtual city' initiatives based on the World Wide Web. Again, the urban analogy is usually employed here to provide the interface between the user and services – a tangible, superimposed image of how the use of electronic spaces relates to the home city of the user. Virtual cities, which again are linked into the City.Net network, often mix urban marketing applications (tourist and conference information), with an integrated 'web' home for any World Wide Web server that falls within the city boundary. Many virtual cities are also developing as part of broader regional strategies aimed at boosting educational, cultural, economic and social development of regions through broad-based telematics-based strategies (Graham and Aungi, 1997). The various European-backed initiatives within the 'Tele-regions' and 'Inter-Regional Information Society Initiative' (IRISI) are examples of these broad strategies aimed at developing telematics cultures in cities and regions.

A particularly interesting and complex virtual city is the *Digitale Stad* (DDS or 'Digital City') initiative in Amsterdam which is one of an integrated network of virtual cities or digital towns in the Netherlands. Funded by Amsterdam City Council, the Ministry of Economic Affairs, Dutch Telecom and the Ministry for Home Affairs, DDS defines itself as a 'test bed where the roots of electronic community can grow' (Digitale Stad, 1996). By late 1994, the system had 7,000 registered users, 120,000 monthly 'hits', and more than a million consulted 'pages' every month. Its main objectives are threefold:

- to widen participation in telematics ('it is of paramount importance that each and everyone wishing to participate in this new digital society is enabled to do so in full measure');
- to develop and disseminate knowledge ('DDS engages in the dissemination of knowledge towards citizens, project partners, providers of services, civic organizations and other digital cities in the Netherlands and in Europe');
- economic development ('the globalization and automatization of the economy are putting employment, and, hence, the cohesion of society under heavy pressure. The digital city wishes to participate in the renewal of small and middle-sized enterprises in the Amsterdam region in order to strengthen its economic structure').

The most interesting point about DDS is that it provides a carefully planned set of virtual spaces intended to act as a powerful metaphor rebounding on to the development of Amsterdam. The spaces within DDS are designed and allocated to reflect directly the sort of convivial urbanism, fine-grained mixture of public, private and domestic space, social democratic values, organic development and mixed-use urban landscapes that Amsterdam embodies physically. Information and communication, and public and private uses, are carefully balanced to try and match that in the city itself. The virtual spaces on DDS are constructed as a series of carefully iconed 'town squares' – meeting points for people sharing an interest in that particular theme (the book square, culture square, gay square, education square, kids' square, technology square, television square, tourist information square, world square, political square, ports square, European square, etc). In a pattern reminiscent of the famed urban layouts of Christaller, every square has four others bordering it, allowing easy web-style exploration. Each square also has its own 'café' or 'pub' where visitors may, within a WWW environment, engage in (archived) discussion in real time about the square's theme. 'News-stands' at each square allow global Internet newsgroups and web sites to be accessed related to the square's themes, so blending local and global. Every square also has a function bar which employs sophisticated search programs to allow visitors to obtain key-word indexed information.

Around each square are eight 'buildings' which are rented to information providers relevant to the themes. One of these is defined as a 'collective building', run by a public or not-for-profit organization. In this way, community centres and clubs could become junctions for community processes and participation. Companies or organizations using these buildings can also advertise on the square's 'billboard'. Clicking the billboard accesses the information pages of that company. Finally, DDS offers residents of Amsterdam the opportunity to build their own 'homes' between the squares, allowing individuals to produce their own information and have electronic meetings.

New Municipal Connections: The Electronic Delivery of Urban Services

Finally, telematics provides new opportunities for changing the ways in which services are delivered to citizens within cities. The information, communications and transactional capabilities of telematics are being explored throughout Western nations as a new means of delivering urban public services (OTA, 1993). Increasingly, the Internet, electronic information kiosks, videotex terminals, automatic teller machines and smart cards are being used to cut the costs and improve the effectiveness of dealing with the millions of day-to-day information requests, communications, and transactions between urban government and citizens. In many American states, for example, government is following the private sector with advanced telematics systems development. These support the electronic delivery of benefits and replace the physical offices of certain government services with electronically mediated kiosks (OTA, 1993).

But such innovations can be shaped in many ways, and the trade-off between risks and benefits remains unclear. On the one hand, they may simply support the substitution of the physical apparatus of urban government with electronic spaces – a way of cutting costs and improving efficiency. In New York, the city's I-Net optic fibre system is already used for conducting remote videoconferences between prison inmates and their legal aid lawyers. 'Touch screens' and kiosks in shopping malls are rapidly emerging in the United States, but with a counter-trend towards the withdrawal of public bureaucracies to fewer back offices (Bellamy and Taylor, 1994). In California and Hawaii, multimedia kiosks in shopping malls offer anything from advice on HIV and local services, through job listings to transactional capabilities such as local fishing or driving licences. But such initiatives may lead to major job-shedding in local government. One company markets them under the heading 'Fewer workers . . .' (North Communications, n.d.). A California policy-maker who uses them admits that 'if hundreds of kiosks are answering routine questions . . . the bureaucracy can function with fewer workers'. An industry spokesman predicts that 'they could eventually replace $6–$8 [per hour] workers' (North Communications, n.d.).

Other emerging examples of the possible substitution of electronic for physical space can be drawn from France. Here the ubiquitous Minitel system is supporting many information, communications and transaction applications linking citizens and municipalities: a theatre booking system operates in Metz; special graffiti removal hotlines have been set up in Bordeaux; part of the Nantes job market has gone on-line; and babysitting brokerage operates in Blagnac. The emergence of hundreds of specialized phone help-lines is also part of this trend.

It seems likely that these processes of change will parallel those in banking and retailing: electronic, home-based services could substitute for physical networks of offices. As well as the likely job losses, they bring with them issues of equity and privacy. How, for example, will people without phones or computer literacy fare in electronic service delivery, especially when their physical access to services may be lost through office restructuring? As with shifts in banking and retailing, these processes may advance the interests of socially privileged, mobile and technologically literate groups while compounding the many disadvantages already faced by marginal groups of 'information have nots' (Dutton, 1993, OTA, 1993). As Dutton (1993: 23) argues, 'living in an information society, the educated public often takes exposure to information technology for granted. Yet information technology is invisible to many within the inner city. . . . In an era of so-called information overload, few managers or professionals can imagine a situation in which there is truly a lack of essential information, but this is precisely the case in the inner city'. Huge demands for information within disadvantaged areas of cities thus often go unmet, a problem that can be compounded by regressive shifts in urban public support services.

On the other hand, though, telematics may be used to radically improve the delivery, user-friendliness and quality of public services to remote locations and people with poor mobility. For example, the kiosks mentioned above may improve the quality and timeliness of information delivery and make citizen–government transactions, in several languages if necessary, easier on both sides. The Santa Monica Public Electronic Network (PEN) – a municipal network aimed at improving citizen–municipal links – 'is widely perceived by city personnel to have enhanced the responsiveness of the city' to citizens (Dutton, 1993: 23). An interactive cable service in Iowa City in the USA offers 70 types of civic information to people's homes (Bankston, 1993). A teleshopping experiment in the 1980s in Gateshead, England, supported the accessing of basic foodstuffs by groups unable to leave their homes (Ducatel, 1994).

The time–space flexibility of telematics may also underpin truly beneficial systems for information exchange, communication and transaction between urban government and its citizens. These may be more appropriate to the wide-ranging needs of the diverse cultural, geographical, ethnic, social and gender groups that make up cities. People with poor physical mobility have potentially most to gain from the use of electronic services to overcome space and time barriers. The resource-sharing potential of telematics may also support radically new innovations in distance learning and education, adult training, community building and democratization in access to information, services and skills. Local cable and telematics systems are being used in several American cities and in Birmingham in the UK to provide special educational services to pre-school children. In New York, a distance learning and videoconferencing network, based on a NYNEX optic fibre infrastructure, offers to the city's schools remote access to the city's cultural and educational institutions.

Conclusions: Grounding the Global: Urban Places, Electronic Spaces

Much popular and media debate on technology and cities implies that cities are merely passive victims of the 'impacts' of telematics. In most scenarios, cities simply serve (implicitly) as the empty arenas where some technological 'revolution' is being enacted. Often, telematics are seen to offer some simple transcendence of the familiar place-based world of urban life – a solution to the perpetual shortcomings, imperfections and finitudes of the urban world. Here, ever more realistic electronic simulations will emerge, allowing geographical place, itself, to be transmitted, accessed remotely and – by implication – made obsolete (Pinney, 1992; Negroponte, 1995). On the other hand, some influential critical literature suggests that urban policy-makers and planners are little more than helpless pawns in the struggle to meet the needs of international capital and international elites.

This discussion leads to three conclusions. The first is that both of these approaches – the technological determinist and the social determinist – are extremely unhelpful (Law and Bijker, 1992: 290). Both, in fact, are short-sighted, and dangerous. Social determinism fails to recognize that a range of local possibilities can be socially constructed in developing contemporary urban policies, a point which has important implications for debates about telematics. They ignore the fact that 'cities can, to some degree, determine their own futures' (Judd and Parkinson, 1990: 29). The more extreme forms of technological determinism and utopianism, meanwhile, in their infectious rush towards 'world rejection' (Schroeder, 1994), serve to deny the continued importance of place, and the politics of place, as crucial factors in contemporary life. Often they ignore the fact that, no matter how 'tele-mediated' urban life becomes (even for the minority groups who dominate telematics use), it will never become truly placeless and free-floating in some electronic ether.

Instead, a complex articulation between life in urban places and interaction in electronic spaces seems to be emerging. 'Constructed spaces', writes William Mitchell (1994: 16), 'will increasingly be seen as electronically-serviced sites where bits meet the body – where digital information is translated into visual, auditory, tactile or otherwise sensorily perceptible form, and vice versa. Displays and sensors for presenting and capturing information will be as essential as doors'. As Robins suggests, 'through the development of new technologies, we are, indeed, more and more open to experiences of de-realization and de-localization. But we continue to have physical and localized existences. We must consider our state of suspension between these conditions' (Robins, 1995: 153). In other words, the contemporary city, while housing vast arrays of telematic 'entrance points' into the burgeoning worlds of electronic spaces, is still a meaningful place economically, socially and culturally (Wilson, 1995; Thrift, 1996).

In fact, cities and telecommunications stand in a state of recursive interaction, shaping each other in complex ways that have a history running back to the days of the origin of the telegraph and telephone (as the continued urban dominance of telecommunications investment and use makes clear; see Thrift, 1996). Major urban places support dense webs of 'co-presence' that cannot – and will not – be mediated by telecommunications (Boden and Molotch, 1994). This is because they are vital supports to high-level business activities in a risky and volatile global economy (Storper, 1997); because the new urban culture relies on them; and because face-to-face social life derives from them.

In this context, cities can be seen to act to minimize space constraints, so overcoming the time constraints caused by distance barriers (which must be overcome by physical movement). Telecommunications, on the other hand, have the opposite effect – they use time constraints to overcome space constraints, through instant flows of electrons and photons across space. Because these two 'roles' are so complementary, future development in Western urban societies seems likely to be about ever more intimate and continuous links between urban place and electronic space, between local and global, between physical daily life and telemediated flow. What this means for cities and urban life, however, is currently very poorly understood (see Mitchell, 1995; Graham and Marvin, 1996).

The implication of this is that both technological and social determinism must be rejected. Theoretical models which conceptualize the 'social' and 'technological' as being caught up in complex and recursive interactions, rather than in separate realms, are required (see, for example, the Actor Network Theory of Latour and Callon: Latour, 1987). The totalizing and reductionist models of city–technology relations, at the root of current debates about the future of cities, need to be replaced. More sensitive frameworks allowing space for contingency and social innovation are required. These must acknowledge the need to link debates about electronic spaces with the lived daily urban experience of the vast majority of people in Western countries. They must incorporate the fundamental indeterminacy of the technological futures of cities, the fact that social action and agency shapes these futures in real places. And they must recognize that complex 'patchworks' of different social/technological innovations and effects are likely to be a key characteristic of this phase of urban policy innovation and experimentation. There is, in short, an 'untidy reality' of diverse, contingent projects across the urban world. These make totalizing and general assumptions about telematics, cyberspace and the future of cities unhelpful.

Secondly, telematics policies grounded in particular towns and cities are playing important roles in shaping the articulation – or, in Robins's words, the 'suspension' – between place-based and electronically mediated realms. They help, quite literally, to 'ground' the globally integrated world of electronic spaces, making them meaningful in real places, real communities, real lives. There is growing evidence that a wide range of local, urban institutions and agents is emerging providing key sites where the potential offered by the technology is being explored, experimented with and socially shaped in diverse and contingent ways. Mitchell calls these policies – which meld physical constructions in urban places and electronic spaces based 'within' telematics networks – 'recombinant architectures' (Mitchell, 1995). In many towns and cities across the advanced industrial world, local policy and planning are significantly influencing this 'state of suspension'. Most cities now attempt to plan, regulate and shape urban place as well as electronic space. Municipal 'recombinant architecture' is moving quickly into urban policy's mainstream.

Here electronic spaces such as hosts, freenets and virtual cities are being used to try and feed back developmental effects into urban places, rather than simply exposing all the elements of the city to the apparently placeless logic of the global explosion of telemediated flows (Virilio, 1987). Such initiatives are helping to 'ground' and 'embed' new technological innovations in real places, rather than just supporting an ever more momentous delocalization through market-driven forces of globalization in the economy, society and culture. These policies are therefore important innovations – passage points, if you like – between lived material life and the disembedded world of telemediated flows. Meaningful 'enclosures' need to be socially constructed in what Castells calls the global 'space of flows' – the panglobal electronic spaces increasingly dominated by massive media corporations and their commodified, capitalized outputs and applications (Castells, 1989). Social need, the particularities of place, freedom of expression and local cultural diversity tend to be squeezed out of the corporate, commodifying logic of this globalization – 'media conglomerates will not fill the vital educational, civic, and cultural needs' of real places and real cities (Grossman, 1995: 25).

Urban telematics policies are, in other words, important in attempting to, in Amin and Thrift's words, 'hold down the global' (Amin and Thrift, 1994: 10). This might be done by maintaining meaningful economic agglomerations, synergies and linkages at the local level; by animating urban cultures; by supporting local democratic debate; and by strengthening and invigorating local webs of social relations in an increasingly globalized world. Urban telematics initiatives are shaping diverse experiences of 'localization' as telematics technologies are used experimentally, in a wide range of ways, locally; they are likely to feed back in complex (and uncertain) ways into the development of places.

Such initiatives also hold important answers to the crucial questions of how we might envisage the 'local' and 'urban' in a globalizing culture, society and economy; how cities might find 'homes' for themselves within a globalizing civil society. The experience of policies at this nexus between urban place and electronic space seems likely to influence much larger questions about the future of cities and urban politics more generally. How, for example, can cities and city regions respond institutionally to the instabilities and volatility caused by globalization? How can urban politics be remade in ways that fight the growing unevenness and fragmentation of urban social and economic life? What does the city and its politics mean within the global shifts now at work?

Complex interactions between global and local urban telematics developments will have a significant role to play in determining whether new, progressive urban visions can be developed that address the context of globalization, fragmentation and polarization (and within which, ironically, telematics are themselves heavily implicated). Equally, local policies may help sustain the public, social face of networks like the Internet, which seems likely to be commercialized through giant media conglomerates and electronic cash systems (Brody, 1995; Grossman, 1995). Without public access areas, civic applications, training and facilities for disadvantaged groups and areas – all developed within 'recombinant architectures' linked with the planning of urban places – electronic spaces will become ever more globalized, commodified and devoid of true public space (Davis, 1992; Schiller, 1994).

Conversely, this is not to imply that city policy-makers have overwhelming power in influencing how telematics is used in cities. Nor is it to romanticize local policy options as panacea solutions to ensure that telematics supports convivial, egalitarian and sustainable telematics innovation in cities. Far from it, in fact. Many difficult problems confront policy-makers in this new area. Whilst local power exists, it is often tightly circumscribed by funding crises and by the need to compete for economic investment in the global 'place market-place'. Technological and regulatory change in telematics often outstrips the ability of local policy-makers to innovate and adapt to changing circumstances, resulting in technological white elephants. Often, initiatives are unsustainable because they fail to meet the actual needs of users, rely too much on long-term subsidy, on a blindly optimistic 'technology push' idea (see Qvortrup et al., 1987; Qvortrup, 1988; Cronberg et al., 1991). Good-quality evaluation of these failures is rare, so often the same pitfalls are fallen into again and again.

The initiatives developed are often relatively insignificant compared to the broader forces of technological change. Urban policy-makers are as prone as anyone to being seduced by notions of futurism and technological determinism (Strover, 1989). What they do can often be criticized as socially regressive, overambitious and based on dubious assumptions about the genuine power of municipal-level policy-making over the global political economic forces of capitalism. Many mistakes result and projects can have unintended side-effects. Finally, telematics are being used in some cases as support for privatization, reducing social services, and enhancing the degree of surveillance and control that dominant local institutions have over the socially powerless (see Graham and Marvin, 1996).

Nevertheless, it is clear that much promising policy innovation and exploration is emerging in which genuine social telematics innovation in cities is producing real benefits. There are many examples where appropriate, low-cost applications of telematics are helping to open up new policy avenues based on inter-urban collaboration as well as on competition, social empowerment and equalization. There are also some genuine attempts to address the social, economic and environmental problems of cities without falling into the traps of techno-hype and the language of the quick technical fix. Such local policies can clearly alter the technological trajectories of places against the backcloth of broader political and economic trends. They can offer more sensitive, positive and appropriate innovations than those of either the hierarchies of the central state or of untrammelled markets.

Thirdly and finally, these policy innovations have implications for how we understand innovations in telematics and how these influence the technological future of urban society. On the one hand, the design and production of technologies is clearly biased. Transnational corporations gain access to optic fibre and private networks; people in disadvantaged ghettos are lucky to access a pay phone. But once technologies are available, political and social struggle and actions can redirect their application and change their effects just as political and social influences can redirect the shaping of urban politics and the built environments of the physical spaces of cities. This means that the effects of telematics in cities can depend heavily on how they are socially and politically constructed. In other words, neither the utopian 'vision of heaven' nor the dystopian 'vision of hell'

(Harrison, 1995) is useful because both imply an either–or future – of all good or all bad. Good and bad implications for cities and urban social life are likely to be mixed up in complex ways. 'The democratising potential of telematics . . . has to be fought for' (Healey et al., 1995: 294) and local strategies seem likely to have a key role here. People may one minute be accessing teleshopping services which may indirectly damage their local high street and switch revenues to distant and unknowable 'back offices'; the next minute they may switch to the local municipal services for applications that actually help bolster their commitment to the place in which they live.

The key question, then, is this: how can sustainable, successful and truly beneficial telematics applications be socially constructed in real world situations and in real cities? How do we translate this technological potential into sustainable applications that actually meet the day-to-day needs and demands of a largely urban society? To quote Dave Spooner (1992: 13), a former technology policy officer for Manchester City Council in the UK, 'the problem is not one of technical development but the lack of appropriate applications. There are hundreds of technical telematics products that have enormous potential for economic [and social] development [in cities], yet they mostly lie on laboratory benches waiting for someone with an application – a classic 'solution looking for a problem'. The implication is that it is social innovation, rather than just technical innovation, that shapes how telematics develops in cities. Specifically urban telematics policies are one small sub-set of these broader social innovations (the others are driven by technological suppliers, telecom operators, user firms, large organizations like universities, utilities and health services, and the myriad of small telematics users).

This point quickly leads to broader normative questions about the sorts of cities 'we' want, how this 'we' might be identified from the disparate groups that make up cities, and the ways in which telematics might help us create those cities (Burstein and Kline, 1995). As William Mitchell (1995: 5) argues:

> the most crucial task before us is not one of putting in place the digital plumbing of broadband communications links and associated electronic appliances (which we will certainly get anyway), nor even of producing electronically deliverable 'content', but rather one of imagining and creating digitally mediated environments for the kinds of lives that we will want to lead.

Useful World Wide Web Addresses

City.Net can be found at http//www.city.net/. Virtual Manchester is at http//www.u-net.com/manchester/. Amsterdam's Digitale Stad is at http//www.dds.nl/. Other city government Web pages are linked from http//rohan.sdsu.edu/infosandiego/examples/citygov/index.html and http//alberti.mit.edu/arch/4.207/anneb/thesis/addresses.html.

References

ADUML (1991) *Plan de développement urbain de la communication*. Lille: Agence de Développement d'Urbanisme de la Métropole Lilloise.
Amin, A. and Thrift, N. (1994) 'Living in the global', in A. Amin and N. Thrift (eds), *Globalization, Institutions and Regional Development in Europe*. Oxford: Oxford University Press. pp. 1–22.
Architectural Design (1995) *Architects in Cyberspace*. London: Architectural Design.
Aufderheide, P. (1992) 'Cable television and the public interest', *Journal of Communication*, 42(1): 52–65.
Bakis, H. (1995) 'Territories and telecommunications – shift of the problematics. From the "structuring" effect to "potential for interaction".' Paper given at the conference on Telecom Tectonics, Lansing, Michigan, March.
Bankston, R. (1993) 'Instant access to city hall: an examination of how local governments use interactive video to reach citizens', in *Multimedia Communications: Forging the Link* (conference report), Chapter 75.
Batty, M. (1990a) 'Invisible cities', *Environment and Planning B*, 17: 127–30.
Batty, M. (1990b) 'Intelligent cities using information networks to gain competitive advantage', *Environment and Planning B: Planning and Design*, 17(2): 247–56.
Bell, D. (1973) *The Coming of Post Industrial Society*. New York: Basic Books.
Bellamy, C. and Taylor, J. (1994) 'Introduction: exploiting IT in public administration – towards the information polity?' *Public Administration*, 72 (Spring): 1–12.
Boden, D. and Molotch, H. (1994) 'The compulsion of proximity', in R. Friedland and D. Boden (eds), *Now/Here Space, Time and Modernity*, Berkeley, CA: University of California Press. pp. 257–86.
Brody, H. (1995) '<Internet@crossroads.$$$>', *Technology Review*, May/June: 25–31.
Brook, J. and Boal, I. (1995) *Resisting the Virtual Life: The Culture and Politics of Information*. San Francisco: City Lights.
Burstein, D. and Kline, D. (1995) *Road Warriors: Dreams and Nightmares along the Information Highway*. Baltimore, MD: Instant Impact.
Castells, M. (1989) *The Informational City: Information Technology, Economic Restructuring and the Urban-Regional Process*, Oxford: Basil Blackwell.
Corey, K. (1987) 'The status of the transactional metropolitan paradigm', in R. Knight and G. Gappert (eds), *Cities of the 21st Century*. London: Sage.
Cronberg, T., Duelund, P., Jensen, O. and Qvortrup, L. (eds) (1991) *Danish Experiments: Social Constructions of Technology*. Copenhagen: New Society Social Science Monographs.
Davis, M. (1990) *City of Quartz: Excavating the Future in Los Angeles*. London: Verso.
Davis, M. (1992) 'Beyond *Blade Runner*: urban control, the ecology of fear', *Open Magazine* (Westfield, NJ).
Digitale Stad (1996) World Wide Web introduction (address on p. 29).
Ducatel, K. (1994) 'Transactional telematics and the city', *Local Government Studies*, 20(1): 60–77.
Dutton, W. (1993) 'Electronic services delivery and the inner city: the risk of benign neglect', Mimeo, School of Communication, University of Southern California, Los Angeles.

Edge, D. (1988) *The Social Shaping of Technology*. Edinburgh: PICT Working Paper no 1, Edinburgh University.
Featherstone, M. and Burrows, R. (eds) (1995) *Cyberpunk/Cyberspace/Cyberbodies*. London: Sage.
Gibson, W. (1984) *Neuromancer*. London: HarperCollins.
Gökalp, I. (1992) 'On the analysis of large technical systems', *Science, Technology and Human Values*, 17(1): 578–87.
Graham, S. (1992) 'Electronic infrastructures and the city: some emerging municipal policy roles in the UK', *Urban Studies*, 29(5): 755–81.
Graham, S. (1994) 'Networking cities: telematics in urban policy – a critical review', *International Journal of Urban and Regional Research*, 18(3): 416–31.
Graham, S. (1995a) 'Cities, nations and communications in the global era: urban telematics policies in France and Britain', *European Planning Studies*, 3(3): 357–80.
Graham, S (1995b) 'Cyberspace and the city', *Town and Country Planning*, 64(8): 198–201.
Graham, S. (1997) 'Networking the city: a comparison of urban telecommunications projects in Britain and France'. Unpublished PhD thesis, University of Manchester.
Graham, S. and Aurigi, A. (1997) 'Virtual cities, social polarization, and the crisis in urban public space', *Journal of Urban Technology*, 4(1): 19–52.
Graham, S. and Marvin, S. (1996) *Telecommunications and the City: Electronic Spaces, Urban Places*. London: Routledge.
Grossman, L. (1995) 'Maintaining diversity in the electronic republic', *Technology Review*, November/December: 23–6.
Harrison, M. (1995) *Visions of Heaven and Hell*. London: Channel Four Television.
Harvey, D. (1989) 'From managerialism to entrepreneurialism: the transformation of urban governance in late capitalism', *Geografisker Annaler*, 71 (series B): 3–17.
Healey, P., Cameron, S., Davoudi, S., Graham, S. and Madani-Pour (eds) (1995) *Managing Cities: The New Urban Context*. Chichester: Wiley.
Hill, S. (1988) *The Tragedy of Technology*. London: Pluto.
IBEX (1991) *Review of Possible Roles of Teleports in Europe, Parts 1, 2 and 3* (report to the European Commission). Brussels: European Commission.
Imrie, R. and Thomas, H. (1993) 'The limits of property-led regeneration', *Environment and Planning C: Government and Policy*, 11: 87–102.
Jowett, G. (1993) 'Urban communication: the city, media and communications policy', in P. Gaunt (ed.), *Beyond Agendas – New Directions in Communications Research*. New York: Greenwood Press. pp. 41–56.
Judd, D. and Parkinson, M. (1990) 'Urban leadership and regeneration', in D. Judd and M. Parkinson (eds), *Leadership and Urban Regeneration: Cities in North America and Europe*. London: Sage. pp. 13–30.
Knox, P. (ed.) (1993) *The Restless Urban Landscape*. Englewood Cliffs, NJ: Prentice Hall.
Latour, B. (1987) *Science in Action: How to Follow Scientists and Engineers through Society*. Milton Keynes: Open University Press.
Law, J. and Bijker, W. (1992) 'Postscript technology, stability and social theory', in W. Bijker and J. Law, *Shaping Technology: Building Society Studies in Sociotechnical Change*. London: MIT Press.
Lyon, D. (1988) *The Political Economy of the Information Society*. Cambridge: Polity Press.
Mandelbaum, S. (1986) 'Cities and communication: the limits of community', *Telecommunications Policy*, June: 132–40.
Mansell, R. (1994) 'Introductory overview', in R. Mansell (ed.), *Management of Information and Communication Technologies*. London: ASLIB. pp. 1–7.

Matthes, S. Alexander (1986) 'Cities and communication: the limits of community', *Telecommunications Policy*, June: 132–40.
Miles, I. and Robins, K. (1992) 'Making sense of information', in K. Robins (ed.), *Understanding Information: Business, Technology, Geography*. London: Belhaven. pp. 1–26.
Mitchell, W. (1994) 'Building the bitsphere, or the kneebone's connected to the I-Bahn', *I.D. Magazine*, November: 13–18.
Mitchell, W. (1995) *City of Bits: Space, Place and the Infobahn*. Cambridge, MA: MIT Press.
Mulgan, G. (1991) *Communication and Control Networks and the New Economies of Communication*. Cambridge: Polity Press.
Murdock, G. (1993) 'Communications and the constitution of modernity', *Media, Culture and Society*, 15: 521–39.
Negroponte, N. (1995) *Being Digital*. London: Hodder & Stoughton.
North Communications (n.d.) *Multimedia Networks* (promotional brochure).
OTA: Office of Technology Assessment (1993) *Making Government Work: Electronic Delivery of Federal Services*, OTA-Tct-578, Washington, DC: Government Printing Office.
Pinney, C. (1992) 'Future travel: anthropology and cultural distance in an age of virtual reality; or, a past seen from a possible future', *Visual Anthropology Review*, 8(1): 38–55.
Qvortrup, L. (1988) 'The challenge of telematics: social experiments, social informatics and orgware architecture', in G. Muskens and J. Gruppelaar (eds), *Global Telecommunications Networks: Strategic Considerations*. Dordrecht: Kleswer.
Qvortrup, L., Ancelin, C., Fawley, J., Hartley, J., Pichault, F. and Pop, P. (eds) (1987) *Social Experiments with Information Technology and the Challenge of Innovation*. Dordrecht: Reidel.
Rheingold, H. (1994) *The Virtual Community*. London: Secker & Warburg.
Richardson, R., Gillespie, A. and Cornford, J. (1994) '*Requiem for the Teleport? The Teleport as a Metropolitan Development and Planning Tool in Western Europe*'. Newcastle Programme on Information and Communications Technologies, Working Paper 17. Newcastle upon Tyne: Centre for Urban and Regional Development Studies, Newcastle University.
Robins, R. (1995) 'Cyberspace and the world we live in', in M. Featherstone and R. Burrows (eds), *Cyberpunk/Cyberspace/Cyberbodies*. London: Sage. pp. 135–56.
Robins, K. and Hepworth, M. (1988) 'Electronic spaces: new technologies and the future of cities', *Futures*, April: 155–76.
Schiller, D. (1994) 'Info highway or corporate monorail?' *Open Magazine*, Pamphlet Series 28.
Schroeder, R. (1994) 'Cyberculture, cyborg post-modernism and the sociology of virtual reality technologies', *Futures*, 26(5): 519–28.
Shearer, D. (1989) 'In search of equal partnerships: prospects for progressive urban policy in the 1990s', in G. Squires (ed.), *Unequal Partnerships: The Political Economy of Urban Redevelopment in Postwar America*. London: Rutgers University Press. pp. 289–307.
Shields, R. (1996) *Cultures of Internet: Virtual Spaces, Real Histories, Living Bodies*. London: Sage.
Sorkin, M. (ed.) (1992) *Variations on a Theme Park*. New York: Hill & Wang.
Spooner, D. (1992) 'The Manchester host computer communications system'. Paper presented at the Centre for Local Economic Strategies on Local Economic Development and Communications Policy, Manchester, 13 May.
Storper, M. (1997) *The Regional World: Territorial Development in a Global Economy*. New York: Guilford.
Strover, S. (1989) 'Telecommunications and economic development: an incipient rhetoric', *Telecommunications Policy*, September: 194–6.

Sussman, G. and Lent, J. (1991) *Transnational Communications: Wiring the Third World*. London: Sage.
Thrift, N. (1993) 'Inhuman geographies: landscapes of speed, light and power', in P. Cloke, M. Doel, D. Matless, M. Phillips and N. Thrift (eds), *Writing the Rural: Five Cultural Geographies*. London: Paul Chapman. pp. 191–232.
Thrift, N. (1996) 'New urban eras and old technological fears: reconfiguring the goodwill of electronic things', *Urban Studies*, 33: 1463–93.
Toffler, A. (1980) *The Third Wave*. New York: Morrow.
Virilio, P. (1987) 'The overexposed city', *Zone*, 1(2): 14–31.
Virilio, P. (1993) 'The third interval: a critical transition', in V. Andermatt-Conley (ed.), *Rethinking Technologies*. London: University of Minnesota Press. pp. 3–10.
Warren, R. (1989) 'Telematics and urban life', *Journal of Urban Affairs*, 11(4): 339–46.
Webber, M. (1968) 'The post-city age', *Daedalus*, Autumn: 1381–421.
Weckerle, C. (1991) 'Télématiques, action locale et 'l'espace public', *Espaces et Sociétés*, 62–3: 163–211.
Williams, F. (1983) *The Communications Revolution*. London: Sage.
Wilson, E. (1995) 'The rhetoric of urban space', *New Left Review*, 209: 146–60.
Winner, L. (1978) *Autonomous Technology: Technics Out-of-Control as a Theme in Political Thought*. Cambridge, MA: MIT Press.
Winner, L. (1993) 'Beyond inter-passive media', *Technology Review*, August–September: 69.

2 Foreclosing on the City? The Bad Idea of Virtual Urbanism

Kevin Robins

In the following discussion, I am concerned with how it is that new information and communications technologies, especially in the form of the so-called information highway, have come to be linked to amelioristic social visions. More particularly, what interests me is their association with ideas and ideals about the future of the city (one might almost think that the urban question was coming to be thought of as a technological one). Why and what, I shall ask, is the 'virtual city'? And how should we consider the significance of this technological vision of urbanity?

The first thing is to look at how the idea of virtual urbanity has become possible. I shall begin with the logic of globalization, or, rather, with a particular account of globalization. Armand Mattelart (1995) calls it an 'off-the-peg ideology', the corporate ideology of globalization, which describes what is happening in the world from a very narrow and partial perspective, considerably diminishing its complexity. There are two aspects to this corporate agenda. The first relates the global future almost exclusively to the emergence of a global information economy, to the centrality of information and knowledge work, and to the significance of increasing electronic flows through the world's information and communications networks. The second maintains that, as the sovereignty of national governments becomes undermined through the creation of global markets, cities are coming to assume a new prominence in the world economy, as the hubs and control centres of the new corporate networks. And, thereby, we have the idea of the global city as informational, or virtual, city. This ideology of the information economy and city is one that is being actively promoted by corporate interests and ideologies. It is also an idea that has found its way on to political agendas, where there is a desire to find some accommodation to the new global order. The alliance of corporate and political objectives has given rise to a new politics of optimism (Robins, 1997), inspired by the idea of transition to a new global information era.

I will then consider aspects of what this wishful politics is anticipating from the new order (nowhere, it seems, does anybody expect other than good things from the new information and communications technologies). Here, I shall first be concerned with how the new technologies have reactivated the old ideal of communication, giving rise to the increasingly prevalent belief that communications media can help in the restoration of a sense of community and in the strengthening of participative democracy. This leads on to a discussion of how the 'virtual city' serves as a focus for the technocultural imagination of a 'revolution' in social relations and interaction. Though he is guarded in his judgement, Stephen Graham (1996a) clearly anticipates a new 'spirit of urban conviviality'. 'The net,' he says, 'may help in generating a new public sphere, supporting interaction, debate, new forms of democracy and "cyber cultures" which feed back to support a renaissance in the social and cultural life of cities.' The city of the future is conceived in terms of a new and more comforting technological order, transcending the difficult and recalcitrant disorder of real world urbanism. It involves an astonishing idealization of both communication and communications technologies. Considering what has happened as a consequence of previous 'communications revolutions', what reason is there to believe that anything significantly new obtains with these newest of new technologies? While the rhetoric of the technoculture may be about innovation and the future, the ideas behind the virtual ideology are far from new, having their roots deep in our culture, and particularly in the Enlightenment project to achieve rational order and control across the lifeworld (Penny, 1994). Let us recognize the technological illusion for what it is, and, even more importantly, let us not make the mistake of thinking that this technological agenda has anything to do with what is at issue in contemporary urban culture and urbanism.

Finally, having criticized the urban technoculture on its own terms, I shall provide an alternative perspective from which we might judge the relevance and value of the 'virtual city' agenda. I shall place what is happening to contemporary cities in the context of an alternative account of globalization, emphasizing the mobilization of populations who are forced to congregate in the vast and sprawling mega-cities now growing throughout the world. What is going on in these cities contrasts dramatically with the ideal of community that is evoked in the image of the technocity. These global cities are places of cultural encounter and confrontation. They are spaces of disorder. And for those whose imagination (with its Enlightenment roots) is centred around the values of coherence and order, this kind of urbanism must be deeply problematical. In this context, we may see the ideal of the virtual city as a defensive and protective response: in one respect it is about the denial or disavowal, by technological means, of this chaotic and difficult reality; and, in another, it involves the attempt to sustain or restore the values (coherence, order, community . . .) of the older (European) ideal of urbanism that now seems to be in crisis. The new technological urbanism is in fact a conservative urbanism. It is actually opposed to urban change. Surely now the imperative must be to come to terms with the real world of global urban culture. We must, as Michael Dear (1991) argues, learn to live with fragmentation, complication and difference. For there are possibilities in the new disorder, as well as dangers, and they are possibilities that far outweigh the convenience and security of technocultural order.

Globalization as Virtual Capitalism

Our consideration of the new developments in communications technologies must begin, then, with the broader economic and political transformations occurring at the end of the twentieth century – transformations that are generally referred to in terms of the logic of globalization. As two former members of the Clinton administration put it, in assessing the strategic position of the United States, 'information is the new coin of the international realm, and the United States is better positioned than any other country to multiply the potency of its hard and soft power resources through information' (Nye and Owens, 1996: 35). They are concerned with the relation between information power and global economic (and also military) ascendancy. From a very different and critical perspective, Michael Dawson and John Bellamy Foster (1996: 48–9) suggest that 'the main reason for corporate interest in the information highway lies in the fact that it is seen as opening up vast new markets'; this highway seems to have the potential now for the creation of a 'universal market'. They, too, are inviting us to consider the significance of the new information order – associated with the development of what they call 'virtual capitalism' – in the context of global capital accumulation.

Economic globalization may be considered, from one perspective, as a major transformation in corporate strategy, involving both the organization of production and the exploitation of markets on a world scale. Whilst this, indeed, amounts to a significant development and even transformation, it must also be said that it represents the continuation, and perhaps culmination, of capital's historical aspiration to transcend national boundaries and achieve world-scale advantages. It is associated with accelerating mobility – of capital, of production, of goods, of communications, and also of people – across frontiers and boundaries. World-space has consequently become the arena of capitalist enterprise. As Manuel Castells (1994: 21) puts it, globalization brings into existence 'an economy where capital flows, labour markets, commodity markets, information, raw materials, management and organisation are internationalised and fully interdependent throughout the planet'. 'By global economy,' he adds, 'we mean an economy that works as a unit in real time on a planetary scale.' And the new global corporations that are now emerging on the world scene must develop the organizational and technological capacity to coordinate and manage these 'real-time' operations across the world's key markets.

What is crucial in this respect, it is argued, is the development of information and knowledge resources. The global economy is, from this perspective, an information economy. This new salience of information is held to be the first of two critical innovations – the second we shall consider in the following section – in the changing world economy. Production in the advanced capitalist societies, according to Carnoy et al. (1993: 5) 'shifts from material goods to information-processing activities, fundamentally changing the structure of these societies to favour economic activities that focus on symbol manipulation in the organization of production and in the enhancement of productivity'. As Castells (1994: 21) puts it, 'information becomes the critical raw material of which all social processes and social organisations are made'; what is critical in all economic activities now, he says, is 'the capacity to retrieve, store, process and generate information and knowledge'. The global information economy is 'based less on the location of natural resources, cheap and abundant labour, or even capital stock and more on the capacity to create new knowledge and to apply it rapidly, via information processing and telecommunications, to a wide range of human activities in ever-broadening space and time' (Carnoy et al., 1993: 6). The consequence of this development, it should be noted, is that informational and knowledge workers also appear to be more significant in the global economy. This is a point made by Alvin Toffler in his book *Powershift* (1990), where he argues that the creation of a 'super-symbolic economy' is associated with the formation of a highly educated and skilled 'cognitariat' (1990: 20, 75). Even more influentially, perhaps, Robert Reich – who was formerly Secretary of Labor in the Clinton administration – has described, in *The Work of Nations* (1992), the emergence of a new class of what he calls 'symbolic analysts'. These professionals – in banking, law, engineering, accountancy, advertising, the media – are the crucial problem-solvers and strategic brokers in the new knowledge-based economy. They are the emerging elite of the global information economy.

An absolutely crucial factor in the transnational organization of this global economy has been the development of new information and communications technologies, increasingly integrated now into a global technological infrastructure. It is these new telecommunications and computer networks that make possible the instantaneous, 'real-time' coordination of economic activities, permitting organizational decentralization to coexist with functional coordination and integration. Many commentators have identified the emergence of a new global space of electronic flows, with the new communications networks serving as the main conduits for all economic activities and functions. There are intimations of a world characterized by network transactions, involving a whole array of 'tele-' activities (telework, teleshopping, tele-education, telemedicine, televiewing, etc.). With the development of such electronic networks, William Mitchell (1995: 139–40) argues,

> the necessary connection between buyers and sellers is established not through physical proximity but through logical linkage. It is all done with software and databases. Merchants get to potential customers by accessing lists of electronic addresses; the key to successful marketing is not being in the right neighbourhood

with the right sorts of customers for whom to lay out wares, but . . . having the right lists for sending out advertising.

This growing centrality of electronic transactions underpins the construction of information highways in the world's developed countries. 'So a new logic has emerged,' Mitchell concludes. 'The great power struggle of cyberspace will be over network topology, connectivity, and access – not the geographic borders and chunks of territory that have been fought over in the past' (1995: 151). What he is celebrating is the transcendence of geographical distance, and the creation of a new space of corporate mobility and flexibility (I shall discuss Mitchell's arguments more fully and critically later).

These developments seem – and are made to seem – to be about a great deal more than just the global extension of corporate power. It is claimed that they are inaugurating a new social order in which we shall find emancipation as both consumers and citizens. Bill Gates (1995: 158) anticipates the perfection of market mechanisms (in accordance with the ideal of Adam Smith), and the advent of what he calls 'friction-free capitalism', 'a new world of low-friction, low-overhead capitalism, in which market information will be plentiful and transaction costs low'. 'It will,' he adds, 'be a shopper's heaven.' Nicholas Negroponte (1995: 231) is optimistic about 'the empowering nature of being digital': 'the access, the mobility, and the ability to effect change are,' he believes, 'what will make the future so different from the present.' The discourse is not that of corporate command and control, but of the conviviality of the 'global village', and of the new possibilities it affords for enhancing community and democracy. Global networks, according to Linda Harasim (1994a: 3), 'offer a venue for the global village, a matrix where the world can meet'. 'The network has become,' she says, 'one of the places where people meet to do business, collaborate on a task, solve a problem, organize a project, engage in personal dialogue, or exchange social chitchat. . . . Networlds offer a new place for humans to meet and promise new forms of social discourse and community' (1994a: 17, 34). What is notable and significant is the ease with which anticipations of the virtual marketplace can slip into imaginations of virtual community and democracy, a point to which I shall return.

This, then, represents one perspective on the logic of globalization, linking global transformation to the formation of a new virtual order. It is an approach that is being very actively propounded by corporate interests – in the pages of the *Harvard Business Review* especially and also, in their best-selling books, by the likes of Nicholas Negroponte and Bill Gates. Later I shall suggest how else one might think about contemporary change, but first I want to reflect on the political appeal of this corporate ideology of globalization. In the next section, I consider how this global information scenario has managed to impose itself on the political agenda and how it has become associated with the urban question and with speculation about the future of cities.

Political Accommodation and the Politics of Place

The sovereignty of the nation state has been a fundamental political principle of the modern period. But now, in so far as global enterprise no longer functions in terms of the national economy, but across frontiers, national governments seem increasingly powerless to regulate and control its activities, and this principle has been significantly weakened. 'Does Government Matter?' was the front-page headline of an issue of *Newsweek* magazine (26 June 1995). 'A brash new world economy is shoving the old statist structures aside,' the lead story declared; 'it is private, it is fast-paced and it is, by and large, averse to government meddling.' The corporate ideology of globalization has proclaimed that the nation state is now a historical relic. Business consultant Kenichi Ohmae (1993: 78) makes the point with great clarity. 'The nation state,' he says, 'has become an unnatural, even dysfunctional, unit for organising human activity and managing economic endeavour in a borderless world. It represents no genuine, shared community of economic interests; it defines no meaningful flows of economic activity.' In the face of the new transnational realities of contemporary capitalism, national governments are finding it more difficult to maintain their integrity. And in the face of ideological offensives by the corporate protagonists of a borderless world economy, they have been less confident about what constitutes the 'national interest'.

The new developments in global information and communications networks have also been a significant factor in undermining the sovereignty and authority of national governments. What we have seen in recent years is the decline of national communications systems, operating according to criteria of public service, and the emergence of a new elite of mega-corporations, operating on a global scale, and for profit. Communication is fundamental to global capitalism, and, at the same time, increasingly inimical to the national organization of communication space. In this respect, the Internet and the information highway are of central importance. For, as one IBM executive has put it, this global network is set to become 'the world's largest, deepest, fastest, and most secure market-place' (quoted in McChesney, 1996: 6). As Mark Poster (1995: 84) observes, national governments are confounded by communications cultures that disrespect and disavow national frontiers: 'Nation-states are at a loss when faced with a global communication network. Technology has taken a turn that defines the character of power of modern governments.' The idea of the 'global village', which supposedly signifies the demise of different national societies and cultures, is pre-eminently associated with the 'revolution' in communications media.

In the face of global transformation, the dilemma confronting national politicians is formidable. Must they submit to the dissolution of national integrity? Can national priorities and concerns be defended? Is it possible to achieve some kind of compromise between global and national agendas? In these difficult circumstances, what we have seen is the adaptation of nation states to corporate imperatives and interests. National governments have generally sought some kind of accommodation to what they perceive to be the new reality principle of globalization.

How has this accommodation been possible? Here the perspective and agenda developed by Robert Reich has been highly influential. And the fact that he moved from promoting the ideology of globalization in the business press, to being a member of the Clinton administration, was clearly significant in harnessing corporate and political interests (and not just in the USA). Reich believes that the spread of the global economy now means that it is no longer feasible to conceive of an autonomous national economic entity. With the growing flows of capital and investment across borders, he maintains, it is impossible to identify production or products in national terms. The consequence, in Reich's view, is that there can no longer be such a thing as a national corporation or a national industry; what we now have is a 'global web' of corporate capitalism which belongs to no nation in particular. It follows, says Reich, that national governments should not pursue economic policies on the basis of defending their national interests. Today's politicians must recognize that the 'ubiquitous and irrepressible law of supply and demand no longer respects national borders' (Reich, 1992: 244). In this changing global context, he argues, the role and the responsibility of national governments must shift. The national interest must be sustained by new strategies to position the national labour force within the changing global context. Or, put another way, the crucial task is that of furnishing the labour force which can attract corporate investors and employers in the new world economy.

Reich argues that it is the national governments which manage to produce or attract the most highly trained workers – the 'symbolic analysts' – that will be the ones to succeed in the new international economic order. This is the accommodating role that national governments should assume. Reich is aware that this symbolic-analytical labour force is highly mobile, with a considerable degree of autonomy with respect to particular places. There is a great danger, he recognizes, of this new professional class 'seceding' from the national culture, but he believes that there are also possibilities for the revitalization of their national commitment. Others, however, are prepared to pursue the implications of this politics more radically. Ian Angell (1995: 11, 12) describes this new, globally mobile class as a corps of 'economic mercenaries', and he believes that nation states will constantly face the threat of 'the entrepreneurial and knowledge elite . . . mov[ing] on to more lucrative and agreeable climes, and, in the long-term, leav[ing] that country economically unviable, composed solely of the unproductive masses, sliding inevitably into a vicious circle of decline'. The possibility of a national compromise with the forces of globalization is clearly a difficult and precarious matter.

Whatever the problems, this approach has gained sufficient influence to become a new political orthodoxy. In Britain, the analyses and policies of the Labour Party have fallen in line with this American approach – described as 'market-based new Keynesianism'. Two of the most prominent codifiers of the party's electoral programme acknowledge the profound importance of Reich's diagnosis of today's economic realities, and have described Reich as being 'more responsible than anyone for stressing the primacy of investment in skills in the modern world' (Mandelson and Liddle, 1996: 89). (The other key reference point in this technocratic dimension of Labour policy is, of course, Alvin Toffler.) In the Commission on Social Justice's 1993 report, which elaborated New Labour's thinking, there is recognition of the fact that 'globalisation is constraining the power of individual nation-states' – so much so that world trading in finance and currencies has 'effectively created an international market in government policies' (1993: 64, 65). The Commission discerns some leeway for economic policy at the European level, but the thrust of its advocacy is towards education and training as the means to position Britain in the new global economy. The emphasis is on building the high-quality labour force that is considered to be essential in a fast-changing global context. Tony Blair articulated New Labour's position at his party's conference in Brighton in 1995. 'The knowledge race has begun,' he declared.

> We will never compete on the basis of a low wage, sweat shop economy.... We have just one asset. Our people. Their intelligence. Their potential.... It is as simple as that.... This is hard economics. The more you learn the more you earn. That is your way to do well out of life. Jobs. Growth. The combination of technology and know-how will transform the lives of all of us. Look at industry and business. An oil rig in the Gulf of Mexico has metal fatigue; it can be diagnosed from an office in Aberdeen.... Knowledge is power. Information and opportunity. And technology can make it happen. (Labour Party, 1995)

As yet, the British emphasis is less on success in the global market for knowledge and information professionals, and rather more on the value of education for upgrading the skills and competences of young people. New Labour's policy on the information superhighway (Labour Party, 1995) bears this out, and underscores its faith in education as the prime means for successful accommodation to the demands of a global economy. But, in so far as the party's ambition is to make Britain 'the knowledge capital of Europe', it will inevitably find itself caught up in the global competition that is accelerating in the elite segment of the labour market.

Let us now consider the new political agenda further, for this political accommodation to the forces of globalization also translates into a political-geographical accommodation. If the mobility of global corporations is overwhelming the old national boundaries, this by no means constitutes an unconditional freedom of movement, and corporate interests must clearly make some form of territorial commitment. Nor can the mobility of the new knowledge elite be absolute, and it is evident that they, too, must sustain some kind of allegiance, however temporary, to place. What is happening is that both are choosing to attach themselves, always provisionally, to sub-national territories, that is to say, to cities or city regions. And this, after the growing salience of information and knowledge resources, may be said to constitute a second key innovation in the development of the world economy. 'The more national states fade in their role,' Castells (1994: 23) observes, 'the more cities emerge as a driving force. . . . Thus we are witnessing the renewal of the role of regions and cities as locuses of autonomy and political decision.' Global corporations are seeking to draw on the skills and resources contained within particular cities and regions. And cities and regions have, by the same token, the clear motivation to ensure that they are 'plugged into' the global economy; in the words of business consultant Rosabeth Moss Kanter (1995: 154, 151), they must learn 'how to harness global forces for local advantage', recognizing that success 'will come to those cities, states and regions that do the best job of linking the businesses that operate within them to the global economy'. National governments are becoming aware that it is, in fact, the endowments of their major city regions that will be critical to their success, or survival, in the corporate global order.

In this context, it is crucial that city regions compete to attract the new class of information and knowledge 'mercenaries', and that they invest in the information infrastructure necessary to make themselves control centres in the global network economy. In the new corporate-political orthodoxy, the emphases on the information economy and the privileging of urban economies come together in the idea of the 'information city'. 'Information is the key product of cities,' claims Mitchell Moss (1991: 184–5), and if this is so, then the 'the principal function of major world cities today is to provide access to information users and providers engaged in the provision of advanced business services'. For national governments, it has become imperative to 'contain' such a metropolitan node, with its dense agglomeration of advanced business and financial service industries. Two advocates of urban boosterism make clear what successful cities must do: 'Globalisation of the world economy . . . require[s] cities to make significant cultural changes; to develop a sense of purpose and direction; and to create a sense of community and a quality of life that will attract managers, scientists, technicians and white-collar workers that form the backbone of the international knowledge industries' (Behrman and Rondinelli, 1992: 116). The national response to the logic of globalization has been to foster and support the informational infrastructure, services and skills necessary to sustain such a global city. The idea of the information city – which, on a global scale, would result in the establishment of an extensive urban information network that would actually support the mobility of information capital and labour – has become the core element in the new ideology of globalization.

Through this combined geo-economic and geopolitical dynamic, we are seeing the emergence of a new global system of city or region states. Kenichi Ohmae (1993: 78) regards them as the 'natural economic zones' in the new world economy, supplanting the 'unnatural' and 'dysfunctional' nation states. There is now a growing consensus that, under conditions of global accumulation, these city states should mobilize the new technologies to achieve competitive advantage. Stephen Graham (Chapter 1 in this volume) thinks of it in terms of the apparently liberating possibilities for expressing 'local agency' (what could this be? is the 'local' a coherent entity?). He imagines the mobilization of information and communications technologies in terms of technological initiatives for 'grounding the global' (again, a nice-sounding idea, but what does it really mean?). Ian Angell outlines the same scenario – with reference to what he aptly calls 'corporation states' – though from a rather more fierce political perspective, with neo-liberal emphasis on the inter-urban competition that would inevitably ensue:

> As the nation-state mutates into corporation-states in the new order, the role of each corporation-state is to produce the right people, with the right knowledge and expertise, as the raw material for the global companies that profit from the Information Age, to service these companies, and to provide them with an efficient infrastructure, a minimally regulated market, and a secure, stable and comfortable environment. (Angell, 1995: 12)

Envisaging 'the rise of the New City State at the hub of global electronic and transport networks' (ibid.), Angell believes that these cities will compete vigorously, even ruthlessly, to maintain their advantage. He spells out the real consequences of local agency. Whilst national governments may think that the promotion of global cities is in their interest, he suggests that in reality the consequence of global city formation may be quite the contrary: 'it will be inevitable that nation-states will fragment'. 'What European city,' he asks, 'will be the first to break ranks with the nation-state mentality holding back progress?' Angell is quite frank about the social consequences of this competition between places:

> To protect their wealth, rich areas will . . . undertake a rightsizing strategy, ensuring a high proportion of (wealth generating) knowledge workers to (wealth depleting) service workers. Rich areas have to maintain and expand a critical mass of scientific and technological expertise, and use it to underpin an effective education system to regenerate the resource. These rich areas will reject the liberal attitudes of the present century, as the expanding underclass they are spawning, and the untrained migrants they welcomed previously, are seen increasingly as economic liabilities. (ibid.)

With the disabling of the nation state, what equilibrating mechanism would remain to compensate for the new divisions between social groups and between urban places?

Ideologies of Communication and Community

This reality of division and polarization must surely be seen as integral to the global transformation that is now taking place, though it is scarcely acknowledged in the ideologies of the global information society. If we recognize that this is a political-economic transformation, and that the principles governing an information economy are unlikely to differ from those at work in previous rounds of capitalist accumulation, then we would hardly expect it to be otherwise (we might find Angell's scenario disturbing, but it should not surprise us). But, of course, if this disturbing reality could be denied, perhaps it might be possible to imagine that contemporary change is inaugurating a new and better social era. And isn't it this mechanism of denial that is at work when the political-economic perspective is displaced by a technological one, as seems to be the case in much of the contemporary discourse? The perspective of 'technological revolution' makes it possible to occlude the disturbing realities of contemporary change, and even to believe that what is occurring might actually be contributing to social and political amelioration. In political manifestos now, the harsher realities of division and conflict are being dissolved in the soft-focus rhetoric of social consensus and cohesion. The advocacy of technological culture is linked to ideals of communication and community – the restoration of community through the enhancement of communication – promising an ordered refuge from the disorders of change in the real world.

The technocultural discourse encourages us to think that something new is on the horizon (down 'the road ahead'). But with a little reflection – and unless you are overly impressed just by the new equipment – it should be clear that there is nothing really innovative or revolutionary in these aspirations. What we have is simply the next banal formulation of some very old utopian formulations. Armand Mattelart (1994a) has traced the association of communication, communion and community back to the Saint-Simonians. Communication has been associated with transparency, and thereby with understanding, and consequently with social harmony: the utopia or ideology of communication is about 'bringing people together', and this bringing together, it is assumed, will consolidate the bonds of community. Mattelart (1994b: 36) identifies the emergence of 'the idea of communication as the regulatory principle counteracting the disequilibria of the social order'. Far from being inherently emancipatory, this ideology, or religion, of communication – which has perhaps even taken over from the religion of progress – has, for 200 years, functioned as a means of social order and control.

Communication is held to promote greater social intelligibility, which in turn is supposed to enhance the 'general concord' among people (and peoples). It is, Mattelart says, a perspective in which 'every social problem tends to be formulated as a communications equation' (1994b: 218). And every solution is then formulated in terms of a new communications technological fix. Consider Linda Harasim's (1994b: 34) routine assertion that 'the need for human beings to communicate and develop new tools to do so forms the history of civilization and culture', and her corollary assertion that the new computer communications constitute, in this context, an innovative technological means to 'enable humanity to express itself in new and hopefully better ways'. In the same spirit, another enthusiast argues that 'the global Matrix of interconnected networks facilitates the formation of global matrices of minds' (Quarterman, 1994: 56). Or consider how, from his self-consciously 'postmodernist' perspective, Mark Poster (1995: 82) asserts that it is communicating by computer that is crucial to the second media age; the Internet is significant, he believes, in so far as it is creating new kinds of 'meeting places, work areas and electronic cafés in which [the] vast transmission of images and words becomes places of communicative relation'. And if, in each of these cases, communications technologies are regarded as a social and political panacea, it must be because an inadequacy, or a breakdown, in communication is regarded as the fundamental social problem.

Politicians are echoing the same message. Al Gore (1994: 4) foresees 'a new Athenian Age of democracy', made possible by the 'global information highway', which is a technological system from which 'we will derive robust and sustainable economic progress, strong democracies, better solutions to global and local environmental challenges, improved health care, and – ultimately – a greater sense of shared stewardship of our small planet'. All this will be possible because the new electronic highway will 'allow us to share information, to connect, and to communicate as a global community'; it will 'allow us to exchange ideas within a community and among nations' (ibid.). He imagines the possibility of 'a kind of global conversation in which everyone who wants can have his or her say' (1994: 6). In Britain, New Labour has the selfsame message:

> We stand on the threshold of a revolution as profound as that brought about by the invention of the printing press. New technologies, which enable rapid communication to take place in a myriad of different ways across the globe, and permit information to be provided, sought, and received on a scale hitherto unimaginable, will bring fundamental change to all our lives. (Labour Party, 1995)

We are encouraged to pin our hopes on 'the network of communication links that can enable people to talk together, to see each other, and share images, texts and sounds, wherever they are around the world' (ibid.). Politicians must believe this. Somewhere in their political hearts, they must all think that 'fundamental change' will really come about once we can talk properly with each other and enjoy a big one-world conversation. It is as if the world's problems were really the consequence of a (historical) communications deficit.

With the praise of communication comes the political idealization of community. There are some who imagine virtual communities in a 'postmodern' sense, in terms of a new kind of fluid and flexible association – community as 'the matrix of fragmented identities, each pointing toward the other' (Poster, 1995: 89). But, for the most part, the new technocommunitarianism is conservative and nostalgic: it is about the imagined recovery and restoration of some original bond. Cristina Odone (1995) expresses the new ideology with great clarity. 'The Net,' she says, 'has been cast over that collective space once filled by the family hearth, the churchyard, the village market place.' Network connections 'seem set to rekindle that elusive sense of belonging that lies at the core of the traditional community'. Political visions are spun out in her imagination like candyfloss:

> So the Web could serve as the first building block in the creation of a whole new social solidarity, founded upon cross-cultural, interdisciplinary dialogues and cemented in an 'empowerment' and 'enfranchisement' of marginalised individuals. A brave new world where heart and soul are restored to the body politic by giving voice to the voiceless and public space to the individual. (1995)

The world is conceived in the image of the village pump or the town square. Its problems can then seem as manageable as neighbourhood problems in the local community. Odone says concisely what the techno-futurist Howard Rheingold can only say in a fat book, full of similar communitarian aspirations. In *The Virtual Community* (1994), Rheingold, too, argues that the Internet is the means, not only to recover the sense of community, but also to attain 'true spiritual communion' (for an extended critique of Rheingold's book, see Robins, 1996: ch. 4). There is a true religiosity in this kind of electronic evangelism (in this context, we might recall the Catholic mysticism of Marshall McLuhan).

How, we might ask, could the politicians of the 1990s really avoid being taken over by this new communitarian energy? For they have been struggling against the legacy of Reaganite and Thatcherite economic liberalism, and they are faced with the task of trying to recreate forms of social and political cohesion. To this end they have now turned to the philosophy and principles of communitarianism, particularly as these have been developed and propagated in the work of Amitai Etzioni (1995). Etzioni's influence has been significant both in American politics (with Newt Gingrich particularly) and in what New Labour calls its 'tough and active concept of community'. It is an approach, moreover, that has easily been incorporated into the new technocultural politics of the information highway. Al Gore has been drawn to the argument that the new technologies could foster democratic culture through electronic simulation of the Jeffersonian town hall (see Ogden, 1994), and New Labour is attentive to Gore's involvement in such 'virtual town meetings' (Mandelson and Liddle, 1996: 209). It is not difficult, it seems, to believe that the new information and communications technologies will put citizens (or so-called 'netizens') back 'in touch' with each other. They seem to make community instantly available, now as a service (or a commodity) that can be piped into the electronic home: community, or interactivity, for domestic consumption.

At this point, I can no longer put off the questions that I have been wanting to ask throughout this exposition. The first is a simple and obvious one. It asks about the kind of perspective or theory that is being formulated within the technoculture. How should we evaluate it as a contribution to contemporary debates on social and political philosophy? This is a question that I pose in a critical spirit. For it seems to me that it offers an impoverished vision, one entirely lacking in imagination or ambition. The notion of virtual community is about escape from the real world of difference and disorder into a mythic realm of stability and order. The ideal society is imagined in terms of communicative interaction within communities of affinity, irrespective of physical location (and these may be communities of decentred, fragmented subjectivities, or communities of conservative citizens, who think they have centred and coherent identities). Sivanandan (1996: 9) is right: the community of the Internet 'is a community of interests, not of people' (and 'you need people to make a revolution', he adds – 1996: 10). It is the virtual condition of disembodiment and disembeddedness that sustains this ideal. Whereas embodied and situated existence is about the necessity of living with others – and not just others in one's immediate community – whose existence frequently challenges and confounds our own, virtual culture and its ideology of communication sustains what Chantal Mouffe (1993: 5) has called 'the illusion of consensus and unanimity'. In so doing it disavows – through the new technological means that are now at its disposal – the conflicts and antagonisms of the real world. But what if, as Mouffe argues, these are constitutive aspects of social and political life? What if they are actually the condition of possibility for civic and democratic culture?

A second question follows directly from this: what has this technological vision to do with what is really occurring as a consequence of global transformation? For it seems to me that there is a stunning discrepancy between this ideal of virtual community and democracy, and the life experiences of most people in the world now. For them – that great majority who do not belong to the elite class of symbolic analysts – global transformation has brought profound upheaval and disruption. Theirs has been the experience of being unsettled and uprooted from established communities, compelled to lead 'flexible', precarious, transient lives and lifestyles. This being the case, one might be led to suspect that the ideal of virtual community is, for those who can get in on the virtual life, a refuge from this disturbing reality. Perhaps the technological order provides them with a compensatory world, another world insulated from the real world and its discontents? Perhaps it is only under conditions of virtual existence that the sense of community can be sustained now? Let us now consider each of these questions in turn, and in some detail.

A City without Substance

It all comes together in the city. The new city states are said to be emerging as the information and communication centres of global business. Their elite status is linked to the proliferation of satellite dishes and cable grids. And now they are entering the popular imagination as symbols of the new economic order and of the new lifeworld it is supposedly bringing into being. The information city, the virtual city: this, we are being told, is the city of the future. What will life be like in these new virtual environments? The ideologies of the technoculture tell us to look forward to tele-working and cyber-shopping, to the comforts of virtual community, and to the re-creation of the Athenian agora by electronic means. Everything comes together in what Stephen Graham (1996a) calls this 'convivial' new urbanism. We should scrutinize carefully this idea of the virtual city, to consider how it has become the focus for eager imaginings of technological regeneration and renewal.

The virtual city is presented to us as the future space of communication and community. Stephen Graham (Chapter 1 in this volume) offers a moderate version of this technological wishfulness, suggesting (or as he puts it, 'the theory goes . . .') that virtual technologies 'might generate new spaces for interaction, debate and cultural development which feed back positively to help support a renaissance in urban social and cultural life'. His exposition is reticent – maintaining that there are 'dystopian' as well as 'utopian' possibilities, and that 'electronic spaces' must be related to 'urban places' – but there is the clear hope, as he puts it elsewhere, that the new technologies can 'help forge new models of conviviality and social cohesion within and between cities' (Graham and Marvin, 1996: 237). There is the potential, it is said, 'for social and cultural identity and social experience and interaction to become reconstituted with more freedom from time–space constraints within the theoretically unlimited domains of electronic spaces' (1996: 173). Other advocates of techno-urbanism are more bold still, more intoxicated by the 'freedom' of the virtual life. For them, as Michael Ostwald (1993: 18) formulates it, the real challenge is 'to build a medium through which humans can interact and develop communities without the restrictions of real space'. He identifies a historical trajectory through which the intractable reality of actual urban environments may be surpassed or superseded. 'Over the past 100 years in the developed world,' says Ostwald, 'there has been a continual move away from conventional communal spaces. In the future each person may inhabit a virtual world, while real bodies are limited to an armchair, a house, a street, a suburb, that stretches as far as the eye can see' (1993: 23). Perhaps the new technologies are about to overcome the restrictions of embodied existence? Can we now look forward to the pleasures and possibilities of disembodied virtual urbanism? Why hang on nostalgically to actual cities?

A particularly good example of the ideology of communication and culture in cyberspace is elaborated in William Mitchell's book, *City of Bits* (1995). It is a fine specimen, worth a little scrutiny, because it shows clearly what is at issue in the virtual project (in both its less and its more extreme formulations). Mitchell's argument is simple. First, he makes the (already banal) claim that 'we are all cyborgs now . . . modular, reconfigurable, infinitely extensible cyborg[s]' (pp. 28, 30). We are, he says, disembodied beings, who will live henceforth in 'the incorporeal world of the Net' (p. 10). This being the case, Mitchell continues, we must take seriously the new kinds of community that 'we cyborgs' are implicated in, for 'we are entering the era of the temporary, recombinant, virtual organization' (p. 97). 'Community has come increasingly unglued from geography,' proclaims Mitchell; 'communities increasingly find their common ground in cyberspace rather than on terra firma' (pp. 166, 161). This in turn being the case, there must be serious consequences for real places, and particularly cities, as we have known them. Mitchell thinks we must now 'conceive a new urbanism freed from the constraints of physical space' (p. 115). This he calls the 'City of Bits':

> This will be a city unrooted to any definite spot on the surface of the earth, shaped by connectivity and bandwidth constraints rather than by accessibility and land values, largely asynchronous in its operation, and inhabited by disembodied and fragmented subjects who exist as collections of aliases and agents. Its places will be constructed virtually by software instead of physically from stones and timbers, and they will be connected by logical linkages rather than by doors, passageways, and streets. (1995: 24)

Mitchell comfortably concludes that 'the very idea of the city is challenged and must eventually be reconceived' (p. 107).

'We have reinvented the human habitat,' says Mitchell (1995: 166). Why do I find his assertion so objectionable and problematical? Mitchell believes that 'the most crucial task before us' is that of 'imagining and creating digitally mediated environments for the kinds of lives that we will want to lead and the sorts of communities that we will want to have' (p. 5). (I note that Mitchell's question is the very one that Stephen Graham poses at the end of Chapter 1 of this volume.) Why am I having difficulty in containing myself in the face of this confident virtual triumphalism? Why do I protest about Mitchell's optimistic futurism?

Well, for a start, because there is nothing that is significantly revolutionary in this mission. It merely continues and perpetuates, by other means, the project of urban modernism, which has involved the progressive rationalization and ordering of city cultures. Modern planning was driven by the desire to achieve detachment and distance from the confusing reality of the urban scene. Christine Boyer (1994: 43, 45) refers to the significance of Le Corbusier in this context: 'No longer [for him] could the traditional meandering streets dominate the material order of the city'; there must be a 'transformation of both the city and the region into a coherent and uniform order'. The imperative was to regulate and control the urban space. In the 'structured and utopian whole' of the modern city, 'disorder was replaced by functional order, diversity by serial repetition, and surprise by uniform expectancy' (Boyer, 1994: 46). In this struggle to impose order and coherence, Le Corbusier sought to mobilize the power of rational vision. In his conception of panoramic control, order was associated with visibility and transparency. This was the Radiant City. 'Reviving the late eighteenth-century myth of "transparency", both social and spatial,' Anthony Vidler (1993: 36) observes, 'modernists evoked the picture of a glass city, its buildings invisible and society open. The resulting "space" would be open, infinitely extended, and thereby cleansed of all mental disturbance.' By these means, modernist planners like Le Corbusier worked towards the 'suppression of everything they hated about the city' (1993: 40) (and what they hated tells us a great deal about them). They did not want to be touched by what they perceived as its confusion and disorder.

The idea of the virtual city is also about establishing order and coherence, this time in a substitute, or proxy, electronic space. Virtual technologies are pre-eminently visual technologies, and this electronic order continues to be one that is predicated on panoptic control and on transparency. These are technologies which – in new ways, and perhaps to an unprecedented degree – afford detachment and insulation from the contamination of reality. 'As networks and information appliances deliver expanding ranges of services,' William Mitchell (1995: 100) approvingly notes, 'there will be fewer occasions to go out' (elsewhere (1995: 38) he suggests that we 'might just want to stay well away from dangerous places like battlefields or the South Side of Chicago'). In Celebration City, the new Disney Corporation development in Florida, 'a firehose bandwidth of fibre links every home to the Net, offering a quaintly familiar-sounding list of futuramas: home security, linking each resident to a central monitoring point, interactive banking, voting from home, virtual office, easy access to each other' (Hayman, 1996). A sign of urban things to come?

I consider the accelerating progression of this logic of order and control to be deeply problematical. William Mitchell (1995: 121) can think of nothing better than 'the realization of whizzier World Wide Webs, superMUDs, and other multi-participant, urban-scale immersive spaces'. I am dismayed by the closed and airless world that is invoked in his obsessively technological imagination. But I am also concerned by the logic of order as it is manifested in more moderate technological visions. It is worth considering the approach developed by Stephen Graham in Chapter 1 of this volume, for it is a good expression of the ordering imagination in its more ordinary, and apparently benign, guise. We must come to the point by way of an interesting detour. Graham actually presents his argument in terms of finding a way between, or beyond, techno-pessimism (which he thinks of as being too negative) and techno-optimism (which he might be accused of, and wants to distance himself from, perhaps because of its unworldliness). His way forward is to criticize both perspectives for what he considers to be their inherent determinism. Then he feels free to differentiate his own perspective, against that of the determinists, as one that takes account of 'human agency'. So, 'the effects of telecommunications-based change,' Graham tells us, 'may be altered or reshaped through policy initiatives'. In rejecting determinism, he says, we shall be able to recognize 'the apparent "manoeuvring" space left for local innovation'. Fine, I say, but is the problem really resolved by this? Is it enough just to assert the importance of agency? (It is difficult to imagine who would really deny its importance.) I think not. For this immediately raises the question of what particular human or local agents might stand for;

it pushes us to consider the values and priorities according to which they seek to relate 'electronic spaces' to 'urban places'. And we might then reflect on the values that Graham himself would pursue (if he were to act as a local agent). New technologies, he tells us, should 'help "re-connect" the . . . fragments that increasingly characterize contemporary cities'. What Graham is concerned with is cohesion and coherence, the reintegration of local communities, the overcoming of what he perceives as urban fragmentation. He does seem quite optimistic about the technological possibilities for (proactively) reordering the urban environment. Let me confess to a certain pessimism when I hear this said – it is a pessimism, though, about Graham's idea of urbanity. For what is there of substance or significance in this idea of 'reconnecting the fragments'?

This playing off of pessimism against optimism – or of 'dystopianism' against 'utopianism' (how devalued these terms have become in contemporary discourses) – is beside the point. This is not what is at issue, and we should be clear about what is. Stephen Graham, who is echoing William Mitchell, thinks it is all about 'imagining and creating digitally mediated environments for the kinds of lives that we will want to lead and the sorts of communities that we will want to have'. For Graham and Mitchell alike, making this choice is, apparently, now 'the most crucial task before us'. What I am saying, against them, is that this choice is simply not enough, for it requires us to express our agency only within the meagre and restricted technological terms set by the technoculture. The choice, if there is one, must be on a different basis. Could we not also decide against the virtual option altogether? Might we not decide that this kind of life and this sort of community actually have nothing to do with the city culture and urbanity that we prefer? For the city we desire might be the one described in *The Man Without Qualities*. It is the city of

> irregularity, change, sliding forward, not keeping in step, collisions of things and affairs, and fathomless points of silence in between, of paved ways and wilderness, of one great rhythmic throb and the perpetual discord and dislocation of all opposing rhythms, and as a whole resembl[ing] a seething, bubbling fluid in a vessel consisting of the solid material of buildings, laws, regulations, and historical traditions. (Musil, 1995: 4)

A far cry from the coherence and order of virtual urbanity, Musil's city is one in which the streets pulse with disorderly and chaotic life. Shouldn't cities be like that? How could Mitchell ever think that this idea of the city must be reconceived? (How could he ever believe that 'the crucial thing' is 'simultaneous electronic access to the same information': 1995: 22?) The agenda proposed by Mitchell, and endorsed by Graham, is opposed to the real condition of urban culture. That is what I find so depressing about it.

The idea of the virtual city is inimical to real (yes) urbanity, which must surely be about embodied and situated presence, proximity, contact – what the street has come to stand for. There is nothing significantly innovative in what simply extends the powers of ordering and rationalization. We should consider Richard Sennett's argument, developed most recently by *Flesh and Stone*, that modern planning and modern technologies have conspired 'to free the body from resistance . . . weakening the sense of tactile reality and pacifying the body' (1994: 18, 17); they have achieved a 'disconnection from space' and a 'de-sensitis[ation] in space'. There has been a progressive withdrawal from the urban scene, Sennett argues, a loss of contact with urban culture, an evasion of encounter with the others in the city: it amounts to a fundamental disavowal of urban reality. 'The geography of the modern city,' says Sennett, 'like modern technology, brings to the fore deep-seated problems in Western civilization in imagining spaces for the human body which might make human bodies aware of one another' (1994: 21). The idea of the virtual city is anti-urban in this sense. Paul Virilio (1996: 45) thinks of it in terms of 'loss of the other, the decline of physical presence in favour of an immaterial, phantom presence'. At the same time, he observes, as it has become possible to relate to those at a distance (who can be switched off at will), there is also a disengagement from the disturbing, and demanding, reality of those whose existence is immediate (and who can't be switched off). The possibilities of telepresence and virtual connection may be undissociable from the destruction of what the city means (or has meant). 'And in losing the city,' Virilio concludes, 'we have lost everything' (1996: 52).

What is fundamental to urbanity, I reiterate, is embodied presence and encounter. It is a question, as Olivier Mongin (1995: 102) emphasizes, of both the 'individual body' and the 'collective body' of the city. In considering this question of the body in (and the body of) the city, we confront that of implication and engagement in its difficult reality. While the techno-culture values the comfort and security that can be achieved through virtual activities, I want to put a value on exposure and its discomforts. Consider Sennett's important argument – incomprehensible within the technocultural worldview – about the need for the urban body to be aroused by disturbance. 'For without a disturbed sense of ourselves,' he maintains, 'what will prompt most of us . . . to turn outward toward each other, to experience the Other?' (1994: 374). The experience of pain is integral to urban living: 'it disorients and makes incomplete the self, defeats the desire for coherence; the body accepting pain is ready to become a civic body, sensible to the pain of another person, pains present on the street . . .' (1994: 376). (In this context Pontalis's (1981) observations on the positive significance of 'psychic pain' are also highly pertinent.) Pain is an inalienable aspect of urban experience, precisely because conflict and antagonism are constitutive of urban culture. The crisis of contemporary urbanity is a crisis in dealing with this reality (allied with the fantasy of disavowing it through technological means). What is fundamentally at issue, as Joël Roman (1994: 9, 11) argues, is 'a crisis in our representation of social conflict'; it is the growing inability to imagine the city as a 'structured conflictual space'. In this context, the virtual city project may be seen as deepening the crisis, rather than contributing to its solution (why do they always make the assumption that new technologies only solve problems?).

Real-Time . . . or Byzantine?

I have been considering how the processes of globalization have brought into existence what has been called the information city or the virtual city. I have been concerned with how this dull urban vision has come to figure prominently in both economic and political discourses. What is the basis of the investments that are being made in the idea of virtual urbanism? What is the nature of the problem to which this new kind of urbanism seems to offer a solution? In posing these questions, I have been content, so far, to consider this technological imagination of the city on its own terms. Now it is necessary to change our focus and perspective, to dislocate the technological agenda from its familiar context. Let me now approach the virtual imagination from the point of view of what it denies and disavows in the urban scene.

To this end, we should come back to the meaning of globalization. So far, I have considered the corporate ideology of globalization – with its technocratic emphasis on the new information and communications networks, and its tidy vision of the 'global village' – in order to reflect on the significance and the implications of this particular and partial agenda. But as Doreen Massey (1996: 118) complains, its image of absolute mobility and unbounded space is 'one which ignores gross inequalities of mobility and connection, lines of inequality between both social groups and parts of the world, and it is – relatedly – predominantly a view from the vantage point of a relative elite in First World countries'. We must recognize that, for the vast majority of the world's population, globalization is experienced in terms of destabilization and damage to their way of life. Globalization, which amounts to both the extension and the intensification of capitalist social relations, involves, as William Robinson (1996: 15) maintains, 'breaking up and commodifying non-market spheres of human activity, namely public spheres managed by states, and private spheres linked to community and family units, local and household communities'. (As the new technocultural politics pursues virtual democracy and virtual community, it seems entirely unaware of what is making its compensatory gestures necessary.) These destructive transformations are associated with growing divisions between rich and poor on the global scale, what Robinson describes as a 'global social apartheid' (1996: 25), and amounting to 'a form of permanent structural violence against the world's majority' (1996: 22). Against the ideal of technological transcendence, let us juxtapose Robinson's image of globalization as world war, 'a war of a global rich and powerful minority against the global poor, dispossessed and outcast majority' (1996: 14). And against the vision of a virtual new world order, let us place the alternative possibility that humanity may, rather, be 'entering a period that could well rival the colonial depredations of past centuries' (ibid.). What if this disturbing and disorderly possibility is what globalization is actually about?

This is a brutal reality that should be taken to the virtual banquet, where the conversation is all about the future, the road ahead, the information highway that will lead us to a new world order. The consensus, as I have made clear, is that we should look forward to our coming emancipation through virtual means (and as virtual beings – 'we cyborgs'). But this fascination with the merely technological serves to obscure a more fundamental continuity of economic and political dynamics. It is better to look to the past if we want to understand the meaning of globalization. In its historical context, it is possible to see globalization as a continuation, and escalation, of the long-standing objectives and imperatives of capitalist accumulation. Thus, the Midnight Notes Collective (1990) describes contemporary transformations as the 'New Enclosures', aiming to establish their continuity with the 'Old Enclosures' (beginning in the late fifteenth century), which, through a ruthless process of expropriation and uprooting, created the 'free' working population necessary to the 'original accumulation' of capital. Now, at the end of the twentieth century, they argue, 'in the biggest diaspora of the century, on every continent millions are being uprooted from their land, their jobs, their homes through wars, famines, plagues, and the IMF ordered devaluations (the four knights of the modern apocalypse) and scattered to the corners of the globe' (1990: 1). Let us situate 'postmodern virtualities' in this overaching and overshadowing context. And let us also acknowledge that virtual technologies may themselves be implicated in this long historical process of enclosure. Think of the 'information revolution', that is to say, in terms of the 'new information enclosures', and be alert to the ways in which 'the new technologies of the virtual life are set to compound the old system of domination with fresh colonizations' (Boal, 1995: 13).

What kind of city can we imagine in this global context? Of course, it will not be the information city, which is the virtual successor to the Radiant City. This global city is the place where the newly mobilized and displaced populations gather in their millions. Each city contains within itself the dynamics of the new world disorder – its dramatic contrasts of rich and poor, its polarizations and segregations, and its encounters and confrontations. It also constitutes a new kind of city, as the coherent and ordered structure of the 'modern city' becomes overwhelmed and superseded by the sprawling, chaotic mega-city or megalopolis (the information and communications systems are ensnared and entangled in this urban anarchy). Michael Dear and Steven Flusty (1998: 66) describe the urban scene of the mega-city – in their case it is Los Angeles – thus:

> Conventional city form, Chicago-style, is sacrificed in favour of a non-contiguous collage of parcellised, consumption-oriented landscapes devoid of conventional centres yet wired into electronic propinquity and nominally unified by the mythologies of the disinformation sewerway. . . . The consequent urban aggregate is characterised by acute fragmentation and specialisation – a partitioned gaming board subject to perverse laws and peculiarly discrete, disjointed urban outcomes. Given the pervasive presence of crime, corruption, and violence in the global city,

not to mention geopolitically as the traditions of the nation state give way to micro-nationalisms and trans-national mafias, the city as gaming board seems an especially appropriate twenty-first century successor to the concentrically-ringed city of the early twentieth.

This they call the 'spatial logic of keno kapitalism'. Its new kinds of mixture and permutation, combined with perverse energies, are bringing about the dissolution of previous models of urbanism. Finally, let us note that the global city is precisely that: global. As the 'exploding cities' of the Third World proliferate at a much faster rate than those of the First (Linden, 1996), perhaps the distinction between 'modern' and 'undeveloped' urbanism is being eroded? Mike Davis (1992: 20) observes how Los Angeles has now come to resemble São Paolo and Mexico City. In the new global context, the privileged reference point of modern (as Western) urbanism and urban culture is confounded.

Cities are changing, and so is the meaning of urbanism. The protagonist of Juan Goytisolo's novel, *Landscapes after the Battle* (1987) perceives it, and revels in it:

> he savors the fluid, permanent presence of the crowd, its chaotic Brownian movements, its feverish diaspora toward every point of the compass. . . . The complexity of the urban environment – that dense and ever-changing territory irreducible to logic and to programming – invites him on every hand to ever-shifting itineraries that weave and unweave themselves, a Penelope tapestry, a mysterious lesson in topography. (1987: 85–6)

He is in his element in that

> non-aseptic, unsanitized medina where the street is the medium and the vital element, a stage-set peopled with figures and signs where barbarians and helots, foreigners and natives joyously take over the space granted their bodies in order to weave an endless, inexhaustible network of exuberant encounters and untrammeled desires; for the emergence, in the perfectly ordered Cartesian perspectives of Baron Haussmann, of bits and pieces of Tlemcen and Dakar, Cairo and Karachi, Bamako and Calcutta; for a Berlin-Kreuzberg that is already an Istanbul-on-the-Spree and for a New York colonized by Puerto Ricans and Jamaicans; for a future Moscow of Uzbeks and Chinese and a Barcelona of Tagalogs and blacks, able to recite from memory, with an ineffable accent, the verses of the Ode to Catalonia. (1987: 120)

Goytisolo's narrator acknowledges the 'irreversible contamination' of the city; 'the gradual de-Europeanization of the city – the appearance of Oriental souks and hammams, peddlers of African totems and necklaces, graffiti in Arabic and Turkish – fills him with rejoicing' (1987: 85–6). And he recognizes that the transformations which have ensued 'are little by little tracing a map of the future bastard metropolis that at the same time will be the map of his own life' (1987: 86).

Time is also of consequence. 'How,' asks Stephen Graham (1996b: 31), 'can we imagine the "real-time" city?' He thinks of the real-time city as the city in which new communications technologies finally overcome what are described as 'time constraints' and 'temporal barriers' (time is thought of as something to be defeated). The entirely unquestioned assumption is that this real-time city would be a good thing. But why, we have to ask, would we ever want to imagine such a possibility? Would this real time not actually be an oppressive and tyrannical time? For, as Paul Virilio (1996: 79) critically observes, it would only serve 'to liquidate the multiplicity of local times'. And local times, in their diversity, constitute an important human resource. 'Cities can be recognised by their pace,' observes Robert Musil in *The Man Without Qualities*, 'just as people can by their walk' (1995: 3). It is only when we consider time as a medium of possibility – rather than as a problem to be dealt with – that we can acknowledge our need for it to be complex, irreducible, varied in its rhythms. In this respect, perhaps all is not yet lost; for even as the technoculture seeks to persuade us of the supposed virtues of real-time living, other things are happening in our cities. The other reality of urban life – that of the 'dense and ever-changing territory irreducible to logic and to programming' – is also cultivating another time. 'The modern megalopolis,' Goytisolo's narrator says, 'is already going by Byzantium time . . .' (1987: 86). Isn't that the kind of time we might be able to do something with? Let the city be the place of elaborate time, of time with arcane spaces, of opaque time and intriguing time. Let it be Byzantine.

The technoculture does not see, or does not choose to see, the possibilities I am suggesting. For it could only see them as confusing, messy, disorderly possibilities. What I am arguing, moreover, is that the technological vision of the virtual city serves to keep them at bay. The technoculture is a culture of denial or disavowal of these disorderly possibilities of contemporary urban reality. What it seeks is the continuation, for it is a conservative rather than a dynamic force, of particular historical values – the community of the pre-modern city, combined with the rational order of urban modernism. And what is apparent now is that these particular values are sustained at the cost of de-realizing urban reality. At a time when there is a gathering consensus – corporate, political, futurological – around this technological agenda, it is important that we open up the debate, to consider a more vital and robust basis to contemporary urbanity. As Olivier Mongin (1995: 24) argues, the present crisis in urbanism – which is occasioned by the death of the 'classic European city', and by the collapse of modernist urban utopias – is 'a crisis of living together (du vivre-ensemble)'.

In contemplating the possibilities that we have before us, we do, of course, have historical reference points – but they must be ones from the disorderly side of our urban past (the side to which Musil is attentive), from the side of urbanity that has been about what Jacques Dewitte (1994: 322) calls 'a certain openness, a disposition to encounter others and to accommodate the unforeseen'. And what we also have are the anti-systemic possibilities that are emerging in the contemporary reality of global cities – possibilities that exist if we are prepared to put our trust in them, if we are prepared for the experience of being changed by this 'creative, bastard mixture' (Goytisolo). Among those who are now considering the possibility spaces of contemporary urbanism, the key distinction is not that between optimists and pessimists, but between those who look to order and those who feel for disorder.

References

Angell, I. (1995) 'Winners and losers in the information age', *LSE Magazine*, 7(1): 10–12.
Behrman, J.N. and Rondinelli, D.A. (1992) 'The cultural imperatives of globalisation: urban economic growth in the 21st century', *Economic Development Quarterly*, 6(2): 115–26.
Boal, I.A. (1995) 'A flow of monsters: Luddism and virtual technologies', in J. Brook and I.A. Boal (eds), *Resisting the Virtual Life: The Culture and Politics of Information*. San Francisco: City Lights. pp. 3–15.
Boyer, M.C. (1994) *The City of Collective Memory*. Cambridge, MA: MIT Press.
Carnoy, M., Castells, M., Cohen, S. and Cardoso, F.H. (1993) *The New Global Economy in the Information Age*. University Park, PA: Pennsylvania State University Press.
Castells, M. (1994) 'European cities, the informational society, and the global economy', *New Left Review*, 204: 18–32.
Commission on Social Justice (1993) *Social Justice in a Changing World*. London: Institute for Public Policy Research.
Davis, M. (1992) *Beyond Blade Runner: Urban Control and the Ecology of Fear*. Westfield, NJ: Open Media.
Dawson, M. and Foster, J.B. (1996) 'Virtual capitalism: the political economy of the information highway', *Monthly Review*, 48(3): 40–58.
Dear, M. (1991) 'The premature demise of postmodern urbanism', *Cultural Anthropology*, 6(4): 538–52.
Dear, M. and Flusty, S. (1998) 'Postmodern urbanism', *Annals of the Association of American Geographers*, 88(1): 50–72.
Dewitte, J. (1994) 'Le bonheur urbain', *Le Messager Européen*, 8: 301–23.
Etzioni, A. (1995) *The Spirit of Community*. London: Fontana.
Gates, B. (1995) *The Road Ahead*. London: Viking.
Gore, A. (1994) 'Forging a new Athenian age of democracy', *Intermedia*, 27(2): 4–7.
Goytisolo, J. (1987) *Landscapes after the Battle*. London: Serpent's Tail.
Graham, S. (1996a) 'Flight to the cyber suburbs', *Guardian*, 18 April.
Graham, S. (1996b) 'Imagining the real-time city: telecommunications, urban paradigms and the future of cities', in S. Westwood and J. Williams (eds), *Imagining Cities*. London: Routledge. pp. 31–49.
Graham, S. and Marvin, S. (1996) *Telecommunications and the City*. London: Routledge.
Harasim, L. (1994a) 'Global networks: an introduction', in L. Harasim (ed.), *Global Networks: Computers and International Communication*. Cambridge, MA: MIT Press. pp. 3–14.
Harasim, L. (1994b) 'Networlds: networks as social space', in L. Harasim (ed.), *Global Networks: Computers and International Communication*. Cambridge, MA: MIT Press. pp. 15–34.
Hayman, S. (1996) 'Two-dimensional living', *Independent on Sunday*, 30 June.
Information Infrastructure Task Force (1993) *The National Information Infrastructure: Agenda for Action*. Washington, DC: NTIA NII Office.

Kanter, R. Moss (1995) 'Thriving locally in the global economy', *Harvard Business Review*, 73: 151–60.
Labour Party (1995) *Information Superhighway*. London: Labour Party.
Linden, E. (1996) 'The exploding cities of the developing world', *Foreign Affairs*, 75(1): 52–65.
McChesney, R.W. (1996) 'The global struggle for democratic communication', *Monthly Review*, 48(3): 1–20.
Mandelson, P. and Liddle, R. (1996) *The Blair Revolution: Can New Labour Deliver?* London: Faber & Faber.
Massey, D. (1996) 'Politicising space and place', *Scottish Geographical Magazine*, 112(2): 117–23.
Mattelart, A. (1994a) *L'Invention de la communication*. Paris: La Découverte.
Mattelart, A. (1994b) *Mapping World Communication: War, Progress, Culture*. Minneapolis: University of Minnesota Press.
Mattelart, A. (1995) 'Les nouveaux scénarios de la communication mondiale', *Le Monde Diplomatique*, August: 24–5.
Midnight Notes Collective (1990) 'Introduction to the new enclosures', in *The New Enclosures*. Jamaica Plain, MA: Midnight Notes. pp. 1–9.
Mitchell, W.J. (1995) *City of Bits: Space, Place, and the Infobahn*. Cambridge, MA: MIT Press.
Mongin, O. (1995) *Vers la Troisième Ville?* Paris: Hachette.
Moss, M.L. (1991) 'The information city in the global economy', in J. Brotchie, M. Batty, P. Hall and P. Newton (eds), *Cities of the 21st Century: New Technologies and Spatial Systems*. Melbourne: Longman Cheshire. pp. 181–9.
Mouffe, C. (1993) *The Return of the Political*. London: Verso.
Musil, R. (1995) [1930] *The Man Without Qualities*, Vol. 1. London: Minerva.
Negroponte, N. (1995) *Being Digital*. London: Hodder & Stoughton.
Nye, J.S. and Owens, W.A. (1996) 'America's information edge', *Foreign Affairs*, 75(2): 20–36.
Odone, C. (1995) 'A patchwork of catholic tastes', *Guardian*, 18 September:
Ogden, M.R. (1994) 'Politics in a parallel universe: is there a future for cyberdemocracy?' *Futures*, 26(7): 713–29.
Ohmae, K. (1993) 'The rise of the region state', *Foreign Affairs*, 72(2): 78–87.
Ostwald, M.J. (1993) 'Virtual urban space', *Transition*, 42: 4–24, 64.
Penny, S. (1994) 'Virtual reality as the completion of the Enlightenment project', in G. Bender and T. Druckrey (eds), *Culture on the Brink: Ideologies of Technology*. Seattle: Bay Press. pp. 231–48.
Pontalis, J.-B. (1981) *Frontiers in Psychoanalysis: Between the Dream and Psychic Pain*. London: Hogarth Press.
Poster, M. (1995) 'Postmodern virtualities', *Body and Society*, 1(3–4): 79–95.
Quarterman, J.S. (1994) 'The global matrix of minds', in L. Harasim (ed.), *Global Networks: Computers and International Communication*. Cambridge, MA: MIT Press. pp. 35–56.
Reich, R. (1992) *The Work of Nations: Preparing Ourselves for 21st Century Capitalism*. New York: Vintage.
Rheingold, H. (1994) *The Virtual Community*. London: Secker & Warburg.
Robins, K. (1996) *Into the Image: Culture and Politics in the Field of Vision*. London: Routledge.
Robins, K. (1997) 'The new communications geography and the politics of optimism', *Soundings*, 5: 191–202.
Robinson, W.I. (1996) 'Globalisation: nine theses on our epoch', *Race and Class*, 38(2): 13–31.
Roman, J. (1994) 'La ville: chronique d'une mort annoncée?', *Esprit*, June: 5–14.
Sennett, R. (1994) *Flesh and Stone: The Body and the City in Western Civilisation*. London: Faber & Faber.

Sivanandan, A. (1996) 'Heresies and prophecies: the social and political fall-out of the technological revolution: an interview', *Race and Class*, 37(4): 1–11.
Toffler, A. (1990) *Powershift: Knowledge, Wealth and Violence at the Edge of the 21st Century*. New York: Bantam Books.
Vidler, A. (1993) 'Bodies in space/subjects in the city: psychopathologies of modern urbanism', *Differences: A Journal of Feminist Cultural Studies*, 5(3): 31–51.
Virilio, P. (1996) *Cybermonde, la politique du pire*. Paris: Textuel.

3 Information and Communications Technologies: Luddism Revisited

Frank Webster

It is now over 10 years since my book, *Information Technology: A Luddite Analysis* (written with Kevin Robins), was first published, and about a decade and a half since it was being researched and written. We have witnessed a great deal of development since that time in the information technology (IT) realm: PCs have become commonplace, Windows 95 and 98 have arrived, CD-ROMs are widely available, BSkyB has made inroads into terrestrial television's territory. People have even stopped talking about IT, preferring the acronym ICT because it encompasses both information and communications technologies.

One cannot be surprised at this. From early days, it was appreciated that the IT 'revolution' was set to be pervasive, continuous, even ordinary. It was actually one of the aims of the book to underline that information technologies were part and parcel of a historical process, uneven but ongoing, a core feature of which was persistent innovation and change. Thus it wasn't so much a matter of rudely 'waking up' to the 'microelectronics revolution', as the then Prime Minister James Callaghan warned in 1979. More important was to realize that this technological spurt was best understood as an integral element of the continuous adaptation and expansion of advanced capitalist forces which were facing a particular set of circumstances. Hence it was largely a matter of enhancing computer communications as corporate institutions spread their reach, of incorporating electronic devices in television and stereo sets as the market for home entertainments burgeoned, of mechanizing and automating processes where competitive edge might be found.

Looking back over the decade, two things seem to me especially striking. First, at what one might call the intellectual level – though this was always much more than a cerebral matter – the ideas of Daniel Bell have prevailed. He had coined the term 'post-industrial' back in the 1960s, and from the outset it was subject to unrelenting and devastating attack from fellow academics, but here we are in the late 1990s and 'post-industrialism', updated now into the concept of an 'information society' (though the rationale and defining characteristics have changed not a jot), is routinely used as a means of designating the current era. Professor Bell has won out:

it is now quite orthodox to conjure, as a matter of unarguable fact, his 'information society' as the knowledge-based, high tech and service-centred world we allegedly inhabit (cf. Feather, 1994; Haywood, 1995). Even sceptical and serious social scientists have succumbed to this terminology (cf. Lyon, 1988; Stehr, 1994; Lash and Urry, 1995). And even leftist accounts of 'Post-Fordism' and 'flexible specialization' testify to the 'revival of post-industrial theory' (Kumar, 1995; and especially Castells, 1996–8).

Secondly, for me it is especially noticeable that, in the present period, we are experiencing the renaissance of one of our prime targets of the early 1980s, namely a utopianism that seizes on the newest technologies to announce a radical break with current arrangements and the coming of all manner of wonderful changes. We had supposed, naively it turns out, that the ordinariness of computerization, as well as the failure of the 'mighty micro' to bring about earth-shattering transformations during the 1980s, would have disabused and deterred the techno-enthusiasts.

Kevin Robins and myself conceived *Information Technology: A Luddite Analysis*, in part at least, as a riposte to a clutter of politicians, journalistic seers and out-and-out futurists who all insisted that the 'mighty micro' was set to change, dramatically, life as we knew it and that, while things might be uncomfortable for a while, the new technologies would deliver utopia – an 'electronic cottage', an 'Athens without Slaves', or even a 'Leisure Society' were variants of the same theme. If we wanted the finest education imaginable, dreamed about the end of arduous manual work, fancied robots to do the housework, or hoped to benefit from the 'wealth of information', then all we had to do was welcome the silicon chip into our lives (Robins and Webster, 1988).

In 1986 we offered a self-consciously Luddite analysis of what was taking place in the IT/information realm as a counter to this sort of presentation, partly to resist the brushing aside of legitimate criticism as Luddite (stupid and reactive, and thus unworthy of consideration), but, more importantly, as a way of showing that technological change needed to be put into social and economic contexts so that one could understand the ways it was shaped by and incorporated given relationships. As we, and, let it be said, a host of others (cf. Woolgar, 1988; Zuckerman, 1988), undertook this task, so was highlighted the untenability of presuming that technology was an autonomous force which *impacted* on society.

Nevertheless, here we are coming towards the end of the millennium having to endure yet another barrage of neologistic chatter, this time about phenomena such as virtual reality, cyberspace and the information superhighway. Nowadays the central subject tends to be computer communications, notably the Internet, but the messianic appeal of technological innovation remains. Commentators no longer find themselves awed by computer numerical control, robotics, and the automated factory; today's techno-boosters consider interactivity, multimedia and global connectivity to be 'the most powerful juggernaut in the history of technology' (Gilder, 1993: 76; cf. Negroponte, 1995). But through the years the frame remains a shared one: technology is a *deus ex machina* that is transforming everything we do and, so long as we embrace it and don't be obstructive, it will take us all the way to nirvana.

This time around, technological utopianism has three distinctive yet related dimensions. In the first of these it coincides with the extraordinary celebration of subjectivism that is such a hallmark of postmodern thinking. Amidst postmodern culture all talk is frenetically relativist: there is no subject, just an infinity of subjects; there is no interpretation, just an unending plurality of interpretations; there is no difference, just an abundance of differences. Here identities and analyses are matters of choices, options, personal dispositions: 'you choose your topic and offer your explanation while I offer mine and she hers . . .'. The latest developments in ICTs strongly reinforce this position, promising 'postmodern virtualities' (Poster, 1995) that are unique and self-generating. This connects with a second dimension, fed by an apparent fusion of biology and machines, which gives rise to cyborgs (cybernetic organisms) that allow us to choose whatever we want to be now that the old certainties (of nature, of sexuality, of bodily constitution) have been undermined. Both dimensions link with a version of the 'new biology' which embraces 'chaos' and 'complexity' theory, the post-Darwinian idea that organisms are at once so astoundingly complicated that they are 'out of control' (Kelly, 1995), yet they are simultaneously and paradoxically self-adjusting by virtue of this very quality of complexity. Right now, the notion of 'spontaneous order' in the biological realm is being transported into the arena of IT and information (which is anyway, as just noted, merging with the biological), and with it the message that life in the 'information age' is blindingly complex because it is all so much a matter of countless and unique individual actions, but yet it all, miracle of miracles, hangs together. This 'control without authority' relies crucially on the presence and rapid development of the Internet since it is the Net which 'channels the messy power of complexity' (Kelly, 1995: 33), transforming the chaos of participation on the giant network into a functioning maelstrom – or, to adopt Kelly's metaphor, creating a self-adapting beehive from the individual actions of millions of bees (cf. Stock, 1993).

Not surprisingly, this vision is presented, in the late 1990s, with abundant acknowledgement of right-thinking environmental concern: indeed the theme borrows heavily from ecological thinking and Gaia's capacity to transform itself in spite of everything haunts such accounts. It is also readily presented as testimony to the desirability of capitalism, for instance when the market system is described as a 'spontaneous self-organizing phenomenon', a 'flexible economic order [which] emerges spontaneously from the chaos of free markets' (Rothschild, 1992: 341, xiv), and when the likes of awestruck *Wired* editor Kevin Kelly (1995: 33) refer to this 'mystery of the Invisible Hand – control without authority'.

This chapter is intended to be more than a retrospect. It is, above all, a comment on the present status of the 'information revolution'. In what follows I will highlight some of the major aspects of the contemporary situation. However, I will return to the early 1980s, not least to spotlight continuities and differences with the present. Here I begin by considering key features of the current milieu in so far as they impinge on informational issues. I want to insist that, just as was the case in the early 1980s, what is needed is an understanding of these contexts if we are to adequately appreciate informational developments. Thereafter I will proceed to discussion of an issue that has attracted a good deal of recent interest, namely the belief that ICTs can aid the development of community relationships, an issue that can extend from re-establishment of an imagined social equilibrium to the hope that computer communications can strengthen participative democracy and even contribute to a more knowledgeable citizenry.

The Contemporary Context

The present situation may be described summarily as the continuation and acceleration of globalization, the persistence of seemingly intractable economic difficulties and accompanying social malaise, ongoing restructuring, a broken labour movement (at least in Britain), and a convergence of political responses to these developments. Informational issues are at the core of much of this and, indeed, their import and character cannot be appreciated if extracted from it.

Globalization is a long-term development, one still far from accomplished, but which accelerated during and since the 1970s. The term signals the growing interdependence and interpenetration of human relations alongside the increasing integration of the world's socio-economic life. Capitalism has pioneered globalization and exported its central tenets of profit-seeking, commodification, and the sanctity of private property. It has proven extraordinarily successful: it has extended its reach across the globe simultaneously with penetrating deep into intimate realms of life. Today, for the first time in the whole of human history, we may say that the entire planet is organized round a single set of economic principles. Such is capitalism's achievement.

This phenomenon, capitalism, while subject to change and historical development, retains its defining forms and guiding principles. Thus, for example, capitalist activities are today at once worldwide (and rapidly extending into hitherto isolated areas such as the former Soviet Union and China) and, at the same time, well able to enter into spheres that were once relatively immune, such as childcare, personal hygiene, and provision of everyday foodstuffs. Moreover, as it has done this, capitalism is bringing the entire world into networks of relationships such that, for example, we may get our coffee from one part of the world, our wines from another, they their television from one region and their clothing from another, all of this conducted by interconnections which integrate the globe. Of course, trade is still predominantly regional, as is production, as one would expect in a world which is vastly unequal in terms of wealth and resources. Hence most activity takes place within and between regional blocs such as Europe and North America. However, the *tendential* process of globalization is clear to all. Bluntly, the trend is towards the world being the context within which relationships are conducted, no matter how localized and particular an individual life may appear to be experienced.

In addition, and crucial to the operation of globalization, is the expansion of transnational corporations (TNCs) that have provided the major foundations of this phenomenon. I would not want to suggest that the TNCs do not have primary national locations or even national focuses of operations – they demonstrably do (Hirst and Thompson, 1996). What I would insist upon, however, is that they have extended their reach across the globe, a process at once extensive and intensive, which both integrates the world economy and reduces the specifically national role of the TNC. The size and scope of TNCs is hard to grasp, but some idea might be gained by noting that, when the wealth of nations and corporations is scaled, TNCs account for half of the largest 100 units. In fact, in financial terms only a couple of dozen countries are bigger than the largest TNC. The likes of General Motors (1993 revenue $134 billion), IBM ($63 billion), Shell ($95 billion) and General Electric ($61 billion) are indeed 'the dominant forces in the world economy' (Dicken, 1992: 49) and they account for as much as 25 per cent of total world production (1992: 48). Moreover, they are themselves highly concentrated, the biggest of the TNCs accounting for the lion's share of activity in any given sector. For instance, Peter Dicken identifies a 'billion dollar club' of just 600 TNCs which supply more than 20 per cent of total industrial and agricultural production in the world's market economies, yet within these giants 'a mere seventy-four TNCs accounted for fifty per cent of the total sales' (1992: 49).

Globalization, in crucial ways operationalized and constructed by transnational corporations, has a number of especially significant features. Prominent amongst these are the following.

Globalization of the Market

This means that the major corporate players now work on the assumption that their markets are worldwide and that these are now open to all economic entities with the resources and will to participate in them. Of course, even nowadays relatively few TNCs operate with a pure global strategy, but this is undoubtedly the direction in which they are moving. Globalization means that markets are today bigger than ever before and that increasingly they are restricted to those with the enormous resources necessary to support a global presence (Barnet and Müller, 1975). Paradoxically, however, markets are in key respects *more fiercely competitive* than previously precisely because they are fought over by giant corporations with the resources to have a global reach. At one time a national market might have been dominated by a local oligopoly, but, over the years, these have been trespassed upon by outsiders (and energetic indigenous corporations have themselves moved outside their home country to attack other markets). These new challengers, in establishing a global presence, are at once bigger and more vulnerable than hitherto. Look where one will and one sees evidence of this process: for instance, the motor industry now operates at a global level, with vehicles being marketed on a world scale, which means that one-time national champions can no longer be secure, a point underlined by the takeover in 1994 of the last major British motor vehicle manufacturer, Rover, by BMW (though Honda already owned 20 per cent of the 'British' outfit, which was then a subsidiary of British Aerospace). Much the same features are manifest in petrochemicals, pharmaceuticals, computers, film, telecommunications equipment and consumer electronics.

It is undeniable that this world market is roughly divisible into three major segments – North America, Europe and South East Asia – since the remainder of the globe offers pretty poor prospects for return on investment, and the major TNCs operate extensively in all three domains. Moreover, noting this broad tripartite division usefully reminds us of something else that globalization of the market means. I refer here to the emergence in little more than a generation of what are today perhaps the archetypical global corporations: Japanese conglomerates which frequently profess to have no national roots (other than in those countries in which they happen to invest). The likes of Hitachi (1993 revenues $69 billion), Matsushita ($61 billion), Toyota ($85 billion) and Sony ($34 billion) have distinctive global strategies for their product ranges. Over the years, in automobiles, consumer electronics and, most recently, information and communications technologies, these have proven to be highly successful. Simultaneously they have contributed to a massive shake-up of established – and expanding – corporate interests in the West. Whether in automobiles, office-equipment, televisions, video or computers, the Japanese challenge has rocked what was, at least for a time, a comparatively settled economic order.

Globalization of production

As corporations are involved in global markets, they must arrange their affairs on a world scale. Global production strategies are a key feature of such a development, TNCs increasingly arranging, for example, to locate their headquarters in New York City, design facilities in Virginia, manufacture in the Far East, assembly perhaps in Dublin, with sales campaigns coordinated from a London office. The inexorable logic of globalization is for TNCs to plan for such strategies in order to maximize their comparative advantage.

This development, as with the globalization of markets, catapults informational issues to the fore, since how else can market strategies and worldwide manufacturing facilities be organized other than with sophisticated information services? The globalization of production also stimulates the growth of what Dicken (1992: 5) calls 'circulation activities' which 'connect the various parts of the production system together'. That is, an essential condition of the globalization of production has been the globalization of information services, such as advertising, banking, insurance and consultancy services, which provide 'an emerging global infrastructure' (ibid). For instance, American Express, Citicorp, BankAmerica, Lloyds and Merrill Lynch straddle the globe, servicing the corporate industrial outfits which they closely parallel in their structures and orientations.

Globalization of finance

This central aspect of globalization, the spread of worldwide informational services such as banks, suggests something of the globalization of finance, but the latter also refers to something much more, nothing less indeed than the development of an increasingly integrated global financial market. With sophisticated ICTs now in place, plus the deregulation of stock markets and the abolition of exchange controls, we have nowadays facilities for the continuous and real-time flow of monetary information, for round-the-clock trading in stocks, bonds and currencies. These developments have enormously increased both the volume and the velocity of international financial transactions, bringing a heightened vulnerability of any national economy to the money markets.

The scale and speed of these informational flows is awesome. For instance, an *Economist* survey (7 October, 1995) of the 'world economy' reports that daily trade in foreign exchange, around $10 billion in 1973, grew sixfold in the following decade, and then between 1983 and 1992 leapt to an astonishing $900 billion. Further, the *Economist* quotes sources in national central banks which 'suggest that daily turnover is [late in 1995] nudging $1.3 trillion' (1995: 12). Will Hutton (1994) adds that foreign exchange turnover now dwarfs the size of national economies and makes trade flows (a traditional method of measuring national economic activity in terms of import and export levels) appear small in comparison.

Globalization of communications

Another dimension of globalization is the spread of communications networks that straddle the globe. Clearly there is a technological dimension to this – satellite systems, telecommunications facilities and the like – but here I would draw attention to the construction of a symbolic environment that reaches right around the globe and is organized, in very large part, by media TNCs.

For instance, in 1994 a core of movies, originating in the United States, achieved far and away the largest audiences wherever they were shown across the globe. *The Lion King*, *Forrest Gump*, *The Flintstones*, *Mrs Doubtfire* and *Maverick* were box-office leaders in Germany, Britain, Italy, France, Spain and Australia, as well as the USA. This provides audiences, widely diverse in their responses and dispositions though they may be, with a mutual symbolic sphere – and much the same could be said about today's television shows, news agencies, music and, indeed, fashion industries. Whatever qualifications we might want to make about the complexities of reception of these messages, it is hard not to see them, in the round, as necessary elements for the extension of marketing (Mattelart, 1991; Schiller, 1984).

Each of these dimensions of globalization requires and contributes to an information infrastructure to cope with the changed stresses and strains of worldwide operation. That is, as globalization grew and as it continues, so ways of handling information and information flows have been put in place (cf. Castells, 1996). We can identify major elements of this informational infrastructure.

- The worldwide spread and expansion of services such as banking, finance, insurance and advertising are essential components of globalization. Without these services TNCs would be incapable of operation. Information is their business, the key ingredient of their work: information about markets, customers, regions, economies, risks, investment patterns, taxation systems and so forth. These services garner information and they also generate and distribute it, having added value by analysis, timeliness of response or collation.
- Globalization requires the construction and, where necessary, enhancement of computer and communications technologies. In recent years we have seen the rapid installation and innovation of information technologies – from facsimile machines to international computer networks – which are a requisite of coordination of global enterprises.
- This information infrastructure has resulted in the growth of information flows at a quite extraordinary rate. For instance, business magazine *Fortune* (13 December 1993: 37) reports that international telephone connections to and from the United States grew 500 per cent between 1981 and 1991 (from $500 million to $2.5 billion). Elsewhere, there has been an astounding expansion of financial traffic along the international information highways. Exchange rate trading, direct foreign investment patterns, and the markets in bonds and equities have expanded apace, underlining the import in global markets of the flows of financial information.

Political Strategies

The globalization of capitalism and the associated growth of transnational corporations have been beset by persistent and seemingly intractable economic difficulties, social instability, and ever more hectic restructuring so that competitive edge and economic expansion might be achieved. Most of this stems from the fact that, in the words of historian Eric Hobsbawm (1994b: 572), the 'world economy [is] an increasingly powerful and uncontrolled engine'. Throughout advanced capitalist societies in Europe and

America there has been a dramatic fall in manufacturing employment, a tidal wave of takeovers and mergers (which almost always lead to layoffs), and a marked shift towards 'flexible' employment, especially in the service industries, that has been disproportionately associated with the feminization of the labour force. Prospects for attaining secure work have markedly decreased for the majority, with even the highly qualified looking to temporary contracts and part-time employment, a staggering 25 per cent of young males in Britain unemployed, and even successful high tech corporations experiencing 'jobless growth', which combines soaring stock prices and sustained 'downsizing' (Burstein and Kline, 1995).

This scarcity of work, the lack of tenure even when a job is obtained (Hutton, 1995), and the uncertainties induced by participation in a world economy, contribute to a widely observed anxiety and apprehension in everyday life. This has been exacerbated, for very many people, by a redistribution of income away from the poorer groups towards the wealthiest (cf. Rowntree Foundation, 1995). In the United States it has been calculated that average weekly earnings fell by 18 per cent between 1973 and 1995 for most workers, while the remuneration of corporate chief executives went up 66 per cent after tax between 1979 and 1989 alone (Head, 1996; cf. Galbraith, 1992). A similar, if less marked, pattern has been observed in Britain, with the richest 10 per cent increasing their real income by over 60 per cent between 1979 and 1991, while the bottom 10 per cent experienced a 17 per cent fall in real income over the same period (Family Expenditure Survey, 1994). These circumstances have – though this is impossible to demonstrate in precise causal terms – contributed to increases in criminality, and a perceived growth of incivility in many spheres of life (Hobsbawm, 1994a).

An important element of these changes has been the weakening of opposition from organized labour. In Britain a 200 per cent increase in unemployment between 1979 and 1981 – which cut a swathe through manufacturing industry, in which were located the most unionized of workers – assisted in this project. But at least as important was legislation which removed unions' legal rights followed by the crushing blow consequent on the defeat of the miners in the historic 1984–5 confrontation. Since then, in Britain at least, labour groups have been incapable of and unwilling to resist most measures called for by corporate leaders.

Writing in the 1990s it is easy enough to observe that the Thatcherite enthusiasm for *laissez-faire* capitalism has waned. It is a moot point whether it ever persuaded the bulk of the populace, but its failure cannot be a surprise given that, on its own terms, it has failed to resolve major problems. While it reduced inflation, Thatcherism neither secured anything like stable – and still less full – employment, nor did it stem a remorseless growth in crime. Given these circumstances, it was only to be expected that there would be a turning away from political rhetoric of the Right.

However, one legacy that remains from the 1980s is that, to adopt the terminology of Margaret Thatcher, *there is no alternative* (better known by the acronym TINA). The acceptance, across the political spectrum, of market principles (competition, profit-seeking, commodification, private ownership, etc.) is not simply an ideological gift from the one-time Conservative leader. It owes much to the dramatic and irreversible collapse of communism in 1989 and perhaps even more to the inexorable expansion of globalizing tendencies. The former, amounting to nothing less than a *fin de siècle* abandonment of any alternative to capitalism (Fukuyama, 1992), combines with the latter's overweening appetite to expand and thereby place massive constraints on the nation state's ambition to determine its own future.

It has long been axial that the nation state is sovereign and has a right, even a duty, to take control of its destiny. During the twentieth century, with the development of democratic regimes, this has become even more important, with governments charged with carrying out the mandate of the people as decided at the ballot box. This is the aspiration: it is becoming evident that it is unfeasible.

It would be foolish to deny that the nation is important for a great many aspects of life, from law and order to welfare provision, and it remains a crucial component of people's identities, but economically at least it seems increasingly irrelevant. There are two particularly significant indications of this. The first is that the rise to prominence of transnational corporations obscures what is owned by any given nation, making it difficult to designate them as unambiguously British or Japanese. For instance, as early as the 1970s in Britain over 50 per cent of manufacturing capacity in high technology (computers, electronics, etc.) and heavily advertised consumer goods (razors, coffee, cereals, etc.) was accounted for by subsidiaries of foreign firms. Are industries located in this country such as Nissan (Sunderland), IBM (Portsmouth) or Gillette (London), British, Japanese or American? Again, a very large amount, in fact around half, of the top 50 British manufacturing companies' production takes place overseas – something which surely confounds government strategies to bolster 'domestic' industries. A disturbing supplementary question follows: to whom then are these TNCs responsive? If they have substantial investment outside the jurisdiction of what one might think of as their 'state of origin', then to whom are they answerable? That begs the question of ownership, a matter of considerable obscurity, but we can be confident, in these days of global stock market dealings, that TNCs will not be owned solely by citizens of any one nation. To the extent that private corporations remain responsive primarily to their shareholders, then this international ownership necessarily denudes conceptions of the 'national interest' and strategies developed by individual nation states.

A second way in which the nation state is undermined is by pressures generated by operating in a global economic context (Sklair, 1990). If nation states are becoming less relevant to business decisions as investors and TNCs seek the highest possible returns on their capital around the world, then individual countries must encounter overwhelming pressures to participate in, and accord with, the global system. This is nowhere more acutely evident than in the realm of financial flows, with nation states nowadays especially vulnerable as regards currencies and investments should governments attempt to do anything out of line. The integration and interpenetration of global economies has resulted in nations having to shape themselves in accord with international circumstances, the upshot of which is that individual states 'have found it extraordinarily difficult to maintain their integrity in the face of the new international realities of capitalism' (Scott and Storper, 1986: 7).

Most nations now actively seek investment from TNCs, but the necessary precondition of this is subordination to the priorities of corporate interests which are at once committed to market practices (in so far as these maximize their interests) and not restricted to particular territories. Hence the freedom of individual governments to determine their own national policies is inevitably constrained by the need to succour foreign investors.

Again, the outcome of unification of the world's financial markets has been that individual governments find their monetary sovereignty challenged whenever investors and traders sense vacillation or weakness, something experienced in the early 1990s in Britain, Ireland and Spain as well as in other countries when the exchange rate mechanism within the EU was wrecked by foreign currency markets. This means that political options and the autonomy of governments are taken away since an anonymous global capital market rules and its judgements about governments' credit worthiness and sustainability are the ultimate arbiter – and much more important than the opinion of national electorates. It is before these that so many governments quail. If they do not obey the policies that the market approves, then their debt and currencies will be sold – forcing them to face an unwanted policy-tightening (Hutton, 1994).

The business world's journal, the *Economist* (7 October 1995), concedes this readily enough, matter-of-factly asserting that 'nobody disagrees that elected governments now have less control over their economies than they used to', adding that 'as global capital markets continue to expand, so will their power' . . . [and the] price governments have to pay . . . is a weakening of their traditional economic armoury, both in terms of the policy choices available and the effectiveness of the weapons they can use' (1995: 44, 6) For the *Economist* this is all to be welcomed since it injects 'healthy discipline' and, anyway, 'capital markets, driven by the decisions of millions of investors and borrowers, are highly "democratic". They act like a rolling 24-hour opinion poll [and] have much sharper eyes than voters' (1995: 6, 44). Against this, we have the warning of an elected British premier. During the mid-1960s the then Labour Prime Minister Harold Wilson complained of mysterious 'gnomes of Zurich' whose trading in sterling compelled his government to devalue the pound and reduce public expenditure. These experiences are frequently cited as instances of the power of financiers to limit national policies. And so they are, but how much more inhibiting are the pressures of today's immensely more integrated, electronically connected, financial centres.

Though it is easy to understand and sympathize with the political responses to these changed circumstances, they remain remarkable nonetheless. In most political discourse today there is manifest both dissatisfaction with out-and-out Thatcherite ideology and resignation to the determining forces of the global economy. An American, Robert Reich, Secretary of Labor in the Clinton administration during the mid-1990s, best articulates what has emerged as a new political consensus as we approach the end of the millennium. He writes as a self-conscious opponent of Reaganite (read Thatcherite) monetarist economics and as one who is supportive of state intervention, at least for welfare purposes. So he reflects the kinder, more communally concerned, politics of the 1990s. However, Reich also offers an analysis that is startling in its implications for today's national governments. In brief, his argument is that, once upon a time (around about the 1950s), what was good for General Motors and other corporations was good for America since the company's stock was indigenously owned, while production, being centred inside US borders, ensured employment of American workers who were secure and relatively well rewarded. The corollary of this was that politicians, doing what they could to support American corporations, were indeed acting in the national interest since the general public was the beneficiary of continued corporate success whether in terms of dividends, jobs or taxation.

However, times change. Above all, Reich argues, the spread of the global economy means that it is no longer feasible to conceive of an autonomous national economic entity. Reich contends that such is the movement of capital and investment, nowadays crossing borders at a bewildering volume and pace, that it is no longer possible even to identify ownership patterns in terms of nations (cf. Wriston, 1992). With currencies continuously flowing electronically through the world's stock markets, with trading a ceaseless activity that is undertaken at the touch of a few keys of a computer terminal, who now can tell if it is Saudi, British or Philippine shareholders who own Unilever, Du Pont or Nestlé?[1] Anyway, he continues, it does not much matter whence the investment comes since the extension of institutional participation in stock markets means that everyone is involved, if only through his or her pension funds or insurance policies.[2] Furthermore, even products themselves are increasingly difficult to identify as nationally fabricated. Labels apart, Reich contends, production itself is a worldwide activity, with components manufactured in one or more often several places, designed in yet another, and often finally assembled only in a 'country of origin' which means little in the way of domestic jobs and real national contribution to the product. Finally, with production shifting away from *high volume* production towards *high value* products, then the exact location of plant is of vastly diminished significance since the onus is on the knowledge input – which is the greatest cost and most attractive proposition in terms of jobs – to the customized product or service.

The upshot of all this, in the view of Reich (1992), is that there 'is coming to be no such thing as an American corporation or an American industry. The American economy is but a region of the global economy'. What we have instead is a 'global web' of corporate capitalism which belongs to no nation in particular and yet simultaneously belongs to every one of us who pays into an insurance or pension scheme. It follows that American politicians can no longer hanker after economic policies that are premised on their national boundaries since they 'have learned that macroeconomic policy cannot be invoked unilaterally without taking account of the savings that will slosh in and out of the nation as a result'. Today's politicians must recognize that the 'ubiquitous and irrepressible law of supply and demand no longer respects national borders' (Reich, 1992: 243–4).

However, Robert Reich does not despair at this reduction of political options. Transnational corporations might not be beholden to national governments, but since we all have a stakeholding in them this is of no great moment. Moreover, there remains a key role for politicians within the modern nation state. This is the important task of furnishing the labour force which will find employment in the world economy. The government that manages to produce the most appealing workers will find its reward in being the home of the highly paid and hence in charge of an affluent nation. Reich focuses his attention on 'symbolic analysts' who will constitute the employed elite in the globalized market. These are the highly qualified information professionals who are, he believes, the cornerstone of the knowledge-based corporations which now increasingly predominate on the world stage. They are problem-solvers, problem-identifiers and strategic brokers in occupations such as banking, law, engineering, accountancy, advertising and media, and even university professors (1992: 177–8).

Here is a role for national government! It can do nothing about the transformed global economy that has made redundant the notion of a national economy and a nationally based corporation, but it can intervene in the education of its citizens to do its utmost to make them attractive to global companies. If government succeeds in producing 'symbolic analysts' with the four basic skills of abstraction, system thinking, experimentation and collaboration (1992: 229), then it will have a populace that will enjoy the highest-paid jobs in the world (the 'routine production services' and 'in-person services' will go to life's losers who are destined to be manual workers and servants), and government will preside over a wealthy and satisfied people.

Reich is pretty sanguine about America's capacity to produce 'symbolic analysts' since 'no other society prepares its most fortunate young people as well for lifetimes of creative problem-solving, -identifying, and brokering' (1992: 228). However, he is concerned about the widening gap between the capacity of 'symbolic analysts' to prosper in the global economy and the rest of the population (as much as 80 per cent!). Reich favours a range of measures to bind together the nation's citizenry (to which I shall return), but worries about the possibility of 'symbolic analysts' being reluctant to engage with the rest of us. Already he believes they separate themselves in enclaves of housing, schools and recreation, but even more worrying is that, since they are in demand worldwide, why should they remain inside their nation of origin and suffer, say, tax rises to subsidize schools in the inner city ghetto? Might they just follow the global trends and up sticks to those places which best remunerate them?

What is central to Robert Reich's account is the abrogation of politicians' responsibility for the national economy and, by way of a substitute, its identification of a political role in providing an education system up to the job of supplying 'symbolic analysts' attractive to roving capital. The social measures that follow, such as alleviation of class divisions and helping the least fortunate, are all subordinate to this prioritization (Luttwak, 1995).

I am struck by how closely the analyses and policies of New Labour in Britain follow those of their American counterparts in (and without) the Democratic Party. Most recently, for example, two key codifiers of New Labour's electoral programme describe Robert Reich as being 'more responsible than anyone for stressing the primacy of investment in skills in the modern world' because of the 'profound importance' of his diagnosis of today's economic realities (Mandelson and Liddle, 1996: 89). Since his landslide victory in May 1997 Tony Blair is likely to be Prime Minister into the next century, and as such his policy merits attention. In the Commission on Social Justice's report that elaborated New Labour's thinking is the same acknowledgement that 'globalisation is constraining the power of individual nation-states' (1993: 64), so much so that world trading in finance and currencies has 'effectively created an international market in government policies' (1993: 65) that places a strait-jacket on indigenous politics. The Commission does discern some leeway for economic policy at the pan-European level, but the thrust of its advocacy is, at one with its American colleagues, towards education as panacea. As Peter Mandleson and Roger Liddle (1996: 89) candidly put it: 'Governments can best promote economic success by ensuring that their people are equipped with the skills necessary for the modern world'. Here it may not be quite so ambitious as to aspire to create a nation of 'symbolic analysts', but New Labour's vision has the same concern for education as the central mechanism for improving national well-being. With New Labour the emphasis is on 'cumulative learning' (Mandelson and Liddle, 1996: 73) which encourages flexibility and ongoing retraining since these are key qualities required of a fast-changing global economy (cf. Commission on Social Justice, 1994).

Tony Blair voiced New Labour's position at his party's conference in Brighton in 1995:

> Education is the best economic policy there is for a modern economy. And it is in the marriage of education and technology that the future lies. . . . The knowledge race has begun. We will never compete on the basis of a low wage, sweat shop economy. . . . We have just one asset. Our people. Their intelligence. Their potential. . . . It is as simple as that. . . . This is hard economics. The more you learn the more you earn. That is your way to do well out of life. Jobs. Growth.

> The combination of technology and know-how will transform the lives of all of us. Look at industry and business. An oil rig in the Gulf of Mexico has metal fatigue; it can be diagnosed from an office in Aberdeen. European businesses finalising a deal with the Japanese. With simultaneous translation down the phone line, and the calls could even be free. Leisure too. Virtual reality tourism that allows you anywhere in the world. . . . Knowledge is power. Information and opportunity. And technology can make it happen.

It is well known that New Labour places great emphasis on IT, a commitment evidenced dramatically by its agreement with British Telecom to connect free of charge every school, college and library to the network (in return for deregulation in favour of BT). New Labour's policy on the 'Information Superhighway' (Labour Party, 1995) bears this out and underscores its faith in education as the prime means of successful adjustment to the demands of a global economy. Hence the party's ambition to make Britain 'the knowledge capital of Europe' and to avoid, at all costs, it becoming 'an electronic sweatshop' (1995: 6) by improving the citizenry's educational experience (cf. Freeman and Soete, 1994). As if to underline the approach, the most memorable slogan of the 1997 election campaign was Mr Blair's iteration that New Labour's policy was 'education, education, education'.

Familiar Contexts, Derived Ideas

It appears to me crucial, as it did a decade ago, to emphasize that these factors – globalization tendencies, corporate priorities, marketization – provide the determining contexts in the development of ICTs. To be sure, within a good many technological innovations there are genuine opportunities for choosing between significantly different forms of application. But it is a misleading presupposition to start with a given technology and thereafter to imagine all manner of freely chosen possibilities. Far better to comprehend the complex circumstances which have led to the generation of particular technological forms, so that a realistic assessment of their potential uses may be made and real choices made. If we cannot break out of the orthodox frame for approaching technological development, which has it that technology arrives in society from R&D laboratories simultaneously to transform our ways of life and provide us with unconditional options, then we are condemned to wild oscillation between sterile exchanges about technology's profound impacts on society and protestations that every technology allows us to choose what to do with it from infinite possibilities.

It is the real world of global integration, market rivalry and corporate expansion that is driving the 'information revolution', ensuring that computer networks are established which facilitate intra-company information flows and result in an explosive growth of 'information factories' such as

Reuters, Datastream and Dun and Bradstreet, dedicated to servicing business interests and thereby prioritizing corporate clients.[3]

Furthermore, it is essential to appreciate fully that ICTs' development has been, and continues to be, driven by the business priorities of cutting costs, increasing profits, and expanding the value of corporate stock. This has provided an unrelenting pressure towards producing information systems that reduce the need for employees, whether by automating their functions or by minimizing their role and hence cutting their incomes. Amongst social scientists the theme of deskilling has gone out of fashion. Its major proponent Harry Braverman (1974) probably did overstate his case, and he certainly did not prioritize the self-perceptions of employees who may feel they have been promoted because a high tech computer terminal has replaced their typewriter or lathe. Moreover, work is now in such short supply that social scientists perhaps feel that close analysis of how it is constituted is something of a superfluous matter. And yet, in spite of this, the pressure to design work processes to maximize the contribution of technologized information systems and minimize the reliance on employees is unyielding. ICTs, from conception to application, are at the core of this.

For instance, over the past decade management consultants have been ecstatic about the spread of 'lean production' in manufacture and 're-engineering' in services (cf. Womack et al., 1990; Hammer and Champy, 1993). These are processes that have led to massive 'downsizing', increased productivity, and 'jobless growth' in the corporate realm. 'Lean production' is, of course, routinely passed off as an inevitable by-product of the shift to service employment, and hence we are supposed to be not unduly concerned that jobs in engineering, machine tools and product manufacture are disappearing. They were never very appealing anyway, so now little regret is expressed about their reduction and the decline in wages of many manual workers over the past 20 years. The key idea is that, as manufacture becomes 'flexible' and automated, excess workers will drift into service work which is cleaner, more satisfying and requires more skill. But even in services during the 1990s there have been marked job reductions – and the central instrument in this has been ICTs, and it has been ICTs designed to effect that outcome. To be sure, the relative impact of ICTs to date has been less dramatic in the service sector than in manufacture, but as Simon Head (1996: 50) observes, the 'reengineering boom came only in the 1990s when computer software became increasingly able to do the jobs of . . . large numbers of workers in hundreds of companies'. And Head emphasizes that 'the real stars of reengineering are the software and hardware packages that form the core of virtually every reengineering project' (1996: 51). Bluntly, ICT systems are integral to these ongoing processes not because of some fortuitous fit between new technologies and corporate commitments, but because ICTs have been shaped by and put into place with business priorities and interests to the very fore.

It is essential to acknowledge, coldly and soberly, that political strategies which look to reshaping education as the key to escaping the difficulties of the present are resigned to responding to pre-established priorities. The primary role of the world economy and its corporate players in instigating changes in workforces and market conditions means that national governments are facing unrelenting pressure to forfeit their sovereignty. All rhetoric aside, they are not taking control of their destinies; they are acquiescing to powerful imperatives from without.

It follows that, while the global economy is a *fait accompli* with which nations must come to terms, the notion that inside national territories government has meaningful control over educational policy is spurious. Urgent restructuring of schools to upskill pupils, to prioritize particular courses or to redesign the curriculum are not matters of political choice, but are testament to the compulsion of the dynamic internationalized economy to impose its criteria on the nation state (Robins and Webster, 1989).

It is salutary to observe the nature of the jobs which have been generated during the past decade or so, and then to connect these to ambitions to transform education to meet the challenges of the 'information age'. An inordinate amount of newsprint has been taken up by the US's acknowledged success at creating eight million new jobs in recent years. It is true that, amongst these, additional professional jobs have been created, but as a proportion of the whole these are minuscule. Fully 85 per cent of the new jobs generated in the USA are in the lowest-paying industries – the retail trade, personal services and fast food joints (Newman, 1991). They are characteristically casual (in business terminology 'contingent'), insecure, part time and feminized. They require minimal skill levels beyond a cheerful disposition and a willingness to put up with long hours and inadequate income. In addition, in today's automated factories the employees are not expected to have much skill. Top priority is for dexterity, enthusiasm and an ability to work in a team. The fabled 'multiskilled' worker demanded by state-of-the-art industries needs to possess few recognizable skills. Furthermore, in those much more glamorous ICT-using occupations such as are found in service industries, the level of computer skills is, on the whole, elementary, training usually taking no more than a couple of months (Robins and Webster, 1989: 179–98).

Ten years ago there was a good deal of heady talk about 'computer literacy' being essential for our schools and universities if they were to do justice to the career prospects of their charges. A decade of experience with word processors, automated library catalogues, cable services and computer simulation games has surely demonstrated the ease of use of the most advanced technologies, an ease of use emanating from the obvious fact that the most advanced technologies are those which require the least skill to use. In spite of this, in 1995 the Secretary of State for Education and Employment could glibly talk of 'network literacy' as if a capacity to log on to and follow click-and-go menus is something remotely comparable to the capacity to read and write.

The conclusion has to be that investing in education as a means of upskilling a workforce that must come to grips with ICTs is a poor bet –

because the information systems being developed and put in place are so designed as to require little of employees. As the *Economist* (6 April 1996: 23) observes, what industry really needs is 'not more skills, but people able to address an envelope'. Former *Financial Times* journalist Simon Head's words are also worth heeding in this light: 'The economy of lean production and re-engineering has no need for a large and growing supply of young workers who have had the kind of German-style training for technically advanced production work' (1996: 51).

In addition, we ought not to let pass the assertion that the global corporation, at once out of control and yet severely restricting the options of national governments, is characteristically owned by many different nationalities and very many everyday people by virtue of the expansion of institutional shareholding. It is certainly true that it is extremely hard to identify precisely who owns today's transnationals, but it defies credibility to suggest that the sources of control and centres of ownership lie in a global no man's land. IBM is an American corporation, with headquarters in Armonk, New York, where its global strategies are decided upon, and its board is overwhelmingly American. Much of the same, *mutatis mutandis*, goes for the rest of the world's transnational champions – there are few difficulties in defining the national origins of Hitachi, Toshiba, Texaco, General Motors, ICI, Kodak, Siemens and the rest, albeit that all of them will have portions of their stock held by foreign stakeholders. Furthermore, while the spread of pension funds and suchlike has contributed to some depersonalization of ownership, this scarcely amounts to the arrival of 'people's capitalism'. Quite the contrary: in the United States and the UK at least, sustained investigation has revealed that a tiny percentage of people (about 1 per cent of the adult population) enjoys a massive concentration of wealth, control and even political rule (Domhoff, 1979; Useem, 1984; Paxman, 1991; Scott, 1991; Westergaard, 1995). Like it or not, accession to the given structure of globalized capitalism disproportionately benefits a class structure based on privileged ownership and power.

Consensus

I have already mentioned that New Labour has been much influenced by the market-based Keynesianism of Robert Reich, who is also convinced that an education system capable of producing revamped labour skills is the best chance of success in global markets. One can, however, go much further than this to emphasize a more general derivation of Blairite ideas from the United States. In part, perhaps, this reflects admiration of the success of Bill Clinton in winning the 1992 and 1996 presidency, and this undoubtedly promoted the role of the 'spin doctors' in New Labour, but to ascribe it simply to an ambition to win elections in the UK seriously understates the scale and scope of New Labour's borrowing from American thinkers.

There are several themes prominent in New Labour thinking that come from the United States (Wintour, 1995). Most prominent is the conviction that we are entering a new sort of society, one which offers enormous possibilities, but for which we require a new kind of politics. At its core is the concept of the 'information superhighway'.

It is Vice-President Al Gore (1994) who has been most prosaic about the creation of this 'Global Information Infrastructure', and indeed it is he who is credited with coining the term 'information superhighway' (cf. Schiller, 1996: 75–89). Gore foresees 'a new Athenian Age of democracy' built on the 'global information highway', something from which 'we will derive [not only] robust and sustainable economic progress, [but also] strong democracies, better solutions to global and local environmental challenges, improved health care, and – ultimately – a greater sense of shared stewardship of our small planet' (1994: 4; cf. Information Infrastructure Task Force, 1993).

It is this sort of talk, if not quite so grandiose, that characterized Blair's keynote address at the 1995 Brighton conference at which he outlined New Labour's vision. There he gave an impassioned endorsement of the information superhighway not least, it is supposed, to present himself as a Wilson-of-the-Nineties enthusiast for 'modernization' (Webster, 1996). My interest here is not to highlight Blair's debt to American Democrats, noticeable though that is. My chief concern is rather to acknowledge that this derivation indicates a wide-ranging consensus amongst contemporary politicians. On this side of the Atlantic this is evident in the long-term championing of ICTs by Liberal Democrat leader Paddy Ashdown and the early installation by the Conservative government of Kenneth Baker as Minister of Information Technology in 1981 (which he followed with Minister of Education). Across the ocean it is clear from the enthusiasm of Republican Speaker of the House, Newt Gingrich, about the PC, cyberspace, the Internet, 'electronic democracy' and associated phenomena (one dimension of which is Gingrich's conviction that ghetto kids would be better off receiving a laptop rather than a welfare cheque from government), that information is regarded as a bipartisan issue.

Most telling as regards this uniformity of politicians' responses to the new technologies is a shared attraction to and admiration of the writing of futurist Alvin Toffler. The ties between Toffler and Gingrich, between a one-time Marxist who is still left-leaning and an out-and-out neo-Conservative, have often been remarked upon (Grant, 1996). These links can be traced back 20 years and have most recently been evidenced in Gingrich's commendation of Toffler's works to the American public (Reeves, 1995). Less well known, but very revealing, is the opinion of former Labour cabinet minister David Owen (1983), voiced on the centenary of Marx's death, that 'Alvin Toffler has more to offer the Left today than Karl Marx'.

It is in the writing of Alvin Toffler, published in a multimillion-selling trilogy from *Future Shock* (1970), through *The Third Wave* (1980), to

Powershift (1990), that one finds elaborated the conventional wisdom of today's politicians. In easily digested and tersely stated gnomic themes is set out a vision that is etched into the brains of politicians ranging from Margaret Thatcher to Tony Blair, Al Gore to Newt Gingrich. In Toffler's futurism the past is straightforwardly understood: technology has driven the major historical changes and, if this has caused social upheaval for those swamped by innovations, at the end of the day we have all benefited from first the agricultural revolution, later the industrial revolution, and today – and tomorrow – from the 'third wave' revolution being brought about chiefly by the 'information revolution'.

In this unabashed technological determinism all the familiar refrains of the politicians are played loud and long. We are being transformed, willy nilly, by an unstoppable tidal wave that takes us now into a 'super-symbolic economy' (Toffler, 1990: 20) where manual workers vanish, to be replaced by the 'cognitariat' (p. 75) who are highly educated, where companies will be 'essentially non-national' (p. 460) and 'knowledge-based' and infinitely 'flexible', where 'data, information, and knowledge' (p. 240) overturn traditional sources of power and bring forth an individuated, de-massified, and decentralized way of life, and where politicians abandon old-fashioned appeals to mass and class-based actions in favour of a 'mosaic democracy' (p. 251) of coalitions and issues.

One does not need to be a trained textual analyst to see here ideas that are consonant with those of Reich, Blair, et al. And that is my point, since it allows us to appreciate more fully the conformity of politicians' policies and the underpinning of their accounts with a dreary inevitability. To be sure, with Toffler and fellow techno-evangelists (cf. Mitchell, 1995) the new technologies ought to be embraced because they promise to carry us all to salvation. This extreme, even perverse, technological determinism ('you can't stop the tidal wave of technology, but it'll be brilliant if you can just catch it on the cusp and ride along to the "electronic cottage" on that far-off beach') does not find marked favour amongst more sober political practitioners. Nevertheless, their capitulation to the forces of the globalized economy comes down to much the same thing. Since to the politicians this is a new sort of economy that is quite beyond control and quintessentially about computer communications exchanges, information flows, investment in knowledge, and high value rather than high volume manufacture, then the differences between their economic and Toffler's technological determinism are often hard to discern. The political mantra is that we have no alternative but to strive assiduously to ensure that our education systems suit the new global political economy, while the Toffleresque line is that the technological revolution *per se* is sweeping us along towards a knowledge-based economy, come what may. In practice there is little to choose between the two positions.

On top of this it should be appreciated that all of these accounts endorse – indeed generally borrow without acknowledgement – Daniel Bell's theory of 'post-industrialism'. For years now Bell has used the terms 'post-industrialism' and 'information society' as synonyms. It is an extraordinary achievement that he has managed to gain such widespread acceptance for his major concepts, not least since within the professional social science literature he has been systematically critiqued and comprehensively rejected (e.g. Kumar, 1995; Webster, 1995). But in the wider world Bell's terminology and prospectus is adopted without demur. For instance, the G7 leading capitalist nations that have met since 1975 to consider matters of pressing concern, in 1994 urged the development of a 'worldwide information society', to which end they convened a ministerial conference on the subject early in 1995. Elsewhere, New Labour (Labour Party, 1995) can speak unhesitatingly about an 'information society' in which 'knowledge is the source of power'. And it can continue to claim that, since 'the information society will have a beneficial impact on employment' and 'around 80 per cent of all jobs created in Europe in the past five years have been connected with the processing of information', then New Labour's ambition is to develop 'programmes for lifelong learning' that will supply 'the broader education needed to ensure that people not only take advantage of existing innovations, but are also able to contribute to their development, and thus to further job creation'.

Where Daniel Bell is far too sophisticated to endorse the crude technological determinism of Alvin Toffler, New Labour has no such qualms. It is the very fact of new technologies which brings into being the 'information society'. Indeed, ICTs put us 'on the threshold of a revolution as profound as that brought about by the printing press'. New Labour insists that it would like to steer the 'new information society', but at the foundation of its account is the fact that it is new technology which is bringing about 'fundamental change to all our lives' (Labour Party, 1995).

So there we have New Labour's analysis. New technologies are arriving amidst a global economy, and together they herald a beneficent 'information society' – so long as we get Tony Blair at the head of government to smooth out the wrinkles. The concurrence of the 'information superhighway' and the global economy is happenchance, and, anyway, both are 'facts of life' which, while they unavoidably impact on all of us, in themselves are of no great social consequence. The key issue, to New Labour – as to Britain's Liberal Democrats, America's Democrats and, whisper it, William Hague's Conservatives – is to vote the party into office so that it may manage this *fait accompli* most judiciously. Any idea that processes of globalization might involve the further penetration of capitalist relations and criteria, that the spread of the 'information superhighway' might be marked by the priorities of its major, corporate, clients, that the 'information society' concept might be a misrepresentation of the real state of affairs – any such idea apparently never crosses their minds. In much the same way, implications of such contexts for the foreclosure of political options remain unconsidered.

The Return of Community?

The foregoing amounts to a description of the 'new realism' which is a defining feature of the present age. Acknowledging the inevitability of living in an integrated world economy, 'modernizing' left-of-centre as well as right-wing politicians insist that there is no alternative to investing hope in re-gearing the nation's educational resources to correspond to the dictates of information-intensive innovation.

Given such resignation it is disconcerting, to say the least, to come across a spate of unworldly commentary which yet aligns itself with the masters of *realpolitik* (cf. Plant, 1993). This one can only describe as techno-fantasist since it sees in the latest forms of IT a means to deny the compulsions of the external world (Robins, 1995). Evoking the techno-evangelism of Marshall McLuhan, this typically postmodern project sees in 'virtual realities' untold opportunities for rising above mundane reality, of routinely losing and endlessly reconstituting the self (selves) in ways which break out of the prison house of social stereotypes and the fixities of the body (Woolley, 1992). This literature has about it an air of liberationism, a whiff of radicalism with its heady talk of 'pleasure', 'cyberpunk', 'multiple identities' and 'vroomies', a soaring sense of freedom for those 'wired' into 'cyberspace'.

Mark Slouka (1996: 30) appositely depicts this alliance of postmodernism and computer technology as a 'mating of monsters'. When one learns that eight out of ten users of the Internet are well-paid, chiefly male graduates (*Financial Times*, 4 September 1995: 12), and that the majority of devotees of 'artificial reality' belong to a tiny minority of the professionally employed (frequently university faculty), then the eagerness to greet the new technologies as revolutionary breakthroughs in (in)human relationships may begin to be comprehensible, though as a depiction of what is really going on in the world of ICTs it is hopelessly unreliable (McKibbon, 1992; Postman, 1992). When one reads that Intranet facilities, restricted to private organizations, are the fastest-growing elements of the Internet system (*Economist*, 13 January 1996: 73), then these 'virtual reality' enthusiasms are a still more disturbing distortion of reality.

Perhaps more worthy of attention is the re-emergence of the theme of community. At one level this may be interpreted as a facet of Clintonism/Blairism that distinguishes its policies from more right-wing political agendas. That is, while on the economic and technological front there is unanimity across the political spectrum, left-of-centre politics especially is infused with a mission to re-establish community amongst citizens, a notion eminently suited to presentation of a caring and concerned outlook (Sullivan, 1995).

Community is a famously ambiguous concept, but the orientation of 'communitarianism' is clear enough. What its major originator Amitai Etzioni (1995) seeks to build are moral bonds that limit individual excess and unite people. The central mechanisms here are those which prioritize family, neighbourhood and civic relationships, all of which can counter a disproportionate swing towards individualism. The key idea is that communal groups such as family and neighbours exercise a disciplinary role that maintains order which is simultaneously integrative since it provides people with a sense of purpose. In the language of New Labour – which has seized on this theme as it has been revived in the USA – this is a matter of emphasizing the obligations citizens have towards one another and the wider society and, reciprocally, the duty of the state to do all it can to ensure that no one is involuntarily excluded from the wider society. Readers will recognize here a familiar concern of traditional Toryism in Britain: One Nation Conservatism is nowadays presented as One Nation New Labourism.

Anyone familiar with the history of sociological thought will recognize this as more or less directly derivative of Frenchman Emile Durkheim's (1893) classic work on the functions of moral cohesion of a century ago. It is important to restate now what was even then a major criticism of the view that 'community' was the solution to discernible problems of increasing unhappiness, dislocation and criminality. It is unconvincing to contend that proposals to bolster morality can resolve problems – though no doubt they may alleviate them when institutions such as the family are given practical support – which emanate from underlying and ongoing material processes. In other words, how can policies aimed at, say, strengthening neighbourhood ties provide an adequate answer to developments which are impelled by powerful processes that fragment, destabilize and unsettle? Today's globalized capitalism requires mobility of workforces, introduces uncertainty into everyday life, encourages people to pursue hedonistic lifestyles, and stimulates perpetual upheaval (Lasch, 1985). The advanced economies impel dissatisfaction, instability and continuous adaptation, and there is an identifiable affinity between the weakening of communal values and their replacement with an excessively self-centred and culturally relativist (and postmodern) worldview (Lasch, 1995) during the twentieth century. This is something which makes the dream of reconstructing a supposedly lost community hopelessly nostalgic and a diversion from perception of the true dynamics of change.

Attempts to resurrect the spirit of community may sit incongruously alongside social and economic trends which vitiate any such endeavour, but conceptions of community nowadays find expression at quite another level to that of the political. Here they are generally closely connected to ICTs, and particularly the Internet, where many commentators imagine that 'virtual communities', by electronically connecting together people with similar interests, can overcome the isolation of modern life.[4] The reasoning is easy enough to follow: on the Net one can 'reach out and touch' kindred spirits anywhere in the world. Arguments for this range from the rather obvious advantages of e-mail connections to the exotic, where projections of 'virtual reality' technologies suggest total fabrications of relationships, from the cerebral to the sexual, such that you will have no reason to leave your console since it will bring to you all the 'reality' you may decide you would like in 'cyberspace' (cf. Benedikt, 1991; Bukatman, 1993).

Howard Rheingold's *The Virtual Community* (1994) is a good condensation of the pragmatic case for electronic collectivity.[5] Like other virtual communitarians, Rheingold starts out from what he sees as the damaged or decayed state of modern life. The use of computer communications, he argues, is driven by 'the hunger for community that grows in the breasts of people around the world as more and more informal public spaces disappear from our real lives' (Rheingold, 1994: 6). Rheingold emphasizes the social importance of the places in which we gather for conviviality, 'the unacknowledged agorae of modern life'. 'When the automobilecentric, suburban, fast-food, shopping mall way of life eliminated many of these "third places" from traditional towns and cities around the world, the social fabric of existing communities started shredding'. His belief is that virtual technologies can stanch such fragmenting developments. Rheingold suggests that cyberspace can become 'one of the informal public places where people can rebuild the aspects of community that were lost when the malt shop became the mall' (1994: 25–6). In cyberspace, he maintains, we will be able to recapture the sense of a 'social commons'.

The virtual community of the Internet is the focus of a grand project of social revitalization and renewal. Under conditions of virtual existence, it appears possible to recover the values and ideals that have been lost to the real world. Through this new medium, it is claimed, we will be able to construct new sorts of community, linked by commonality of interest and affinity rather than by accidents of location. Rheingold believes that we now have 'access to a tool that could bring conviviality and understanding into our lives and might help revitalise the public sphere'; that, through the construction of an 'electronic agora', we will be in a position to 'revitalise citizen-based democracy' (1994: 14), this time, however, on a global scale.

It is not hard to demonstrate flaws in this argument. Underlying all faults must be objection to the suggestion that technology can create community of itself, that the technical capacity to connect together electronically amounts to the reconstitution of community. Something of the superficiality of this reasoning can be appreciated by asking whether the telephone network, connecting millions of dispersed individuals, operates to unite or rather to maintain the separateness of contemporary life. It is, at the least, arguable that the telephone perpetuates individuated existences since it allows social organization to continue alongside increased privatism. More seriously, the idea that ICTs can re-establish community evokes the obvious, but nevertheless fundamental, criticism that there is no technological route to the reconstruction of lost moral ties between people. Such a project requires reorganization of power structures, restructuring of lifestyles, and reconsideration of the values (or lack of them) by which we live.

Further, the Internet connects only certain dimensions of interacting participants, paying attention only to one's sectional interests and whims. This is surely antithetic to any genuine sense of community in which one encounters the complexities and difficulties involved in relating to whole people, with those features with which one might find sympathy alongside those that irritate, challenge and confound (Talbott, 1995). A cognate problem is that the Internet deals, overwhelmingly, in disparate 'bits' of information, passing across the network enormous quantities of material which encourage users to browse in a random manner, to leap in and out of 'conversations', close off when interest wanes or a disconcerting snippet of information appears.

I want to avoid counterposing 'virtual community' with a conception of genuine community which the Internet can never reproduce. The problem here is that real community – if it ever existed – was rooted in a locality, with routine interpersonal contact, and there can be no return to that. Yet it is a yearning for this lost community which drives the techno-utopians. They search in technology for a comforting, enclosing and assured existence which is escapist. The notion of the 'virtual community' is one of escape from the real world of disorder and difference into a mythic realm of order and stability (Sennett, 1973).

Finally, some attention needs to be paid to the popular notion amongst cybernauts that the network is at once an individualist's delight yet marked by spontaneous social order. Simultaneously appealing to the anarchist streak in many Internet enthusiasts and to the conservative insistence that there must be order at all costs, the theme that there is a network society in which all participants may join together without bowing to an authority structure is especially appealing.

At the start of this chapter I referred to Kevin Kelly's version of this in his book *Out of Control*. George Gilder (1993) sings much the same song when he enthuses that 'computer nets permit peer-to-peer interactivity rather than top-down broadcasts. . . . Computer networks offer as many potential connections as there are machines linked to the web. Rather than a few "stations" spraying images at millions of dumb terminals, in real time, computer networks put the customer in control' (1993: 76). Gilder eagerly looks forward to the collapse of television and its centralized output; it is to be displaced by interactive computer technologies with which individuals take control because they will be able to 'always order exactly what you want when you want it' (1993: 76). Paradoxically, it will be this very diversity introduced by computer technology that brings about spontaneous order since what it results in is 'one colossal processor' to which everyone is connected; everyone does their own thing, but the whole melds together on 'the new worldwide web of glass and light' (1993: 77, 78).

There is more than an echo of Emile Durkheim here too, with a strong evocation of an increasingly individuated society hanging together by virtue of each contributing their specialized part to maintain the organic whole. However, what is still more vividly evoked by this high tech utopia is Friedrich von Hayek's paean to *laissez-faire* capitalism that was published in 1945. In Hayek's terms a complex society renders it impossible for any one single person to know what is required by the rest of us, hence any state attempt to plan centrally is doomed to dissatisfy the populace, and to be inefficient to boot. It follows that the only reliable form of information is that generated by the sovereign individual, by 'decentralisation . . . because only thus can we insure that the knowledge of the particular circumstances of time and place will be promptly used' (Hayek, 1949: 84). In Hayek's view the 'price system' may be regarded as a form of decentralized information assessment. Left alone, the millions of daily information exchanges that are represented by buying and selling ensure that the economy is sensitive to people's desires and, a 'marvel' (*sic*), they ensure social order because 'prices can act to co-ordinate the separate actions of different people' (1949: 85). It is precisely this combination of individualism and order which today's techno-boosters like Gilder and Kelly evoke long after von Hayek spelled it out in his apologia for the market system. Recognition of this lineage helps us appreciate the old refrain in the latest tune.

Conclusion

Information Technology: A Luddite Analysis ended on a gloomy note. Kevin Robins and I had been struck in the early 1980s by the contrast between a rush of technological utopianism and the reality of innovations shaped by, and generally designed to perpetuate and consolidate, an aggrandized system of corporate capitalism. We set out our case, some 200,000 words of empirical and theoretical documentation, to underline the gulf between real social trends and the wishful thinking of many commentators who seemed awestruck by new technologies and the new times they augured, yet simultaneously able to adopt an ultra-realist vocabulary because they told how this was having a devastating impact on our whole way of life.

Arguing like this, I have become used to being pigeon-holed a 'pessimist' who can see no good coming from ICTs (cf. Lyon, 1988; Forester, 1989: 3). Contrasted to my position is that of the 'optimists' who highlight the beneficial dimensions of ICTs. I have to say that I do not find this categorization helpful. It suggests that one's interpretation of what is happening in society is a matter of psychological disposition rather than of substantive analysis. As it happens, I am by nature an optimist. Further, I have been fortunate in my personal circumstances. I sincerely wish that the world would fit with my sunny character and that everyone was as well-favoured as myself. However, and though it has become rather unfashionable to say so in these postmodern times, social reality can be a lot different from one's imaginings. A decade after the publication of *A Luddite Analysis* I can see little reason to cheer up when it comes to assessing what is actually happening in the information domain. To be sure, there are some benefits here and there, and these ought not to be discounted. But in the round what is striking is the contrast between wild optimism about ICTs, and the future in general, and actual empirical evidence of what is happening. Instead of labelling critics 'pessimists', and thereby ignoring their evidence and banishing them from sight, I would like those who disagree with the analysis to provide more adequate empirical investigations. This involves a lot more than waxing poetic about how they are inspired by surfing the Internet or how e-mail gets folk in touch with kindred spirits in California. It requires systematic investigation of the contexts and circumstances shaping the wider information domain.

Because there appears to have been so little resistance to techo-hype now that it has returned once again, perhaps I should be glummer still. As we approach the millennium the roar of inevitability is deafening, the technological determinism as prevalent as ever, the resignation of most participants – bursts of jubilant rhetoric notwithstanding – to received patterns striking.

Notes

1 Actually, government statistics estimate about 12 per cent of shares in UK companies are overseas owned, scarcely amounting to a non-national base (*Social Trends*, 23, 1993: 79, Table 5.22).

2 For instance, in the UK in 1963 personal ownership accounted for 54 per cent of all available shares. However, by 1992 this had fallen to below 20 per cent with pension funds and insurance company holdings accounting for over half (*Social Trends*, 23, 1993: 79 Table 5.22). However, despite privatization programmes throughout the 1980s, in Britain only 20 per cent of the public own shares, and this is highly skewed, with classes A and B (which constitute 18 per cent of the population) owning 33 per cent of all shares (*Social Trends*, 24: 78 Table 5.24).

3 *Business Week* (25 *August* 1986: 50) estimated that almost half of electronic information in the United States is financial information used for transactions such as stock and commodity prices.

4 In keeping with proponents' enthusiasm, much of this finds expression on the Internet rather than in old-world print. Those with access may follow and explore some of this by scanning the 'home pages' of Howard Rheingold or the John Perry Barlow 'Library'.

5 The following draws on Robins (1995).

References

Barnet, R.J. and Müller, R.E. (1975) *Global Reach: The Power of the Multinational Corporations*. London: Cape.

Benedikt, M. (ed.) (1991) *Cyberspace: First Steps*. Cambridge, MA: MIT Press.

Braverman, H. (1974) *Labor and Monopoly Capital: The Degradation of Work in the Twentieth Century*. New York: Monthly Review Press.

Bukatman, S. (1993) *Terminal Identity: The Virtual Subject in Postmodern Science Fiction*. Durham, NC and London: Duke University Press.

Burstein, D. and Kline, D. (1995) *Road Warriors: Dreams and Nightmares along the Information Highway*. New York: Dutton.

Castells, M. (1996–8) *The Information Age: Economy, Society and Culture. Volume 1: The Rise of the Network Society* (1996); Volume 2: *The Power of Identity* (1997); Volume 3: *End of Millennium* (1998). Oxford: Basil Blackwell.

Commission on Social Justice (1993) *Social Justice in a Changing World*. London: Institute for Public Policy Research.

Commission on Social Justice (1994) *Social Justice: Strategies for National Renewal*. London: Vintage.
Dicken, P. (1992) *Global Shift: The Internationalization of Economic Activity*, 2nd edn. London: Paul Chapman.
Domhoff, W.G. (1979) *The Powers That Be: Processes of Ruling-Class Domination in America*. New York: Random House.
Durkheim, E. (1933) [1893] *The Division of Labor in Society*, trans. George Simpson. New York: Free Press, 1933.
Economist (1995) 'The world economy' (survey) 337 (7935), 7 October Inset: 44.
Etzioni, A. (1995) [1993] *The Spirit of Community*. London: Fontana.
Family Expenditure Survey (1994) *Households below Average Income*. London: HMSO.
Feather, J. (1994) *The Information Society: A Study of Continuity and Change*. London: Library Association Publishing.
Forester, T. (ed.) (1989) *Computers in the Human Context*. Oxford: Basil Blackwell.
Freeman, C. and Soete, L. (1994) *Work for All or Mass Unemployment: Computerised Technical Change into the 21st Century*. London: Pinter.
Fukuyama, F. (1992) *The End of History and the Last Man*. London: Hamish Hamilton.
Galbraith, J.K. (1992) *The Culture of Contentment*. London: Sinclair-Stevenson.
Gilder, G. (1993) 'The death of telephony', *Economist*, 328 (7828): 75–8.
Gore, A. (1994) 'Forging a new Athenian Age of democracy', *Intermedia*, 27 (2): 4–7.
Grant, L. (1996) 'Gurus of the Third Wave', *Guardian Weekend*, 13 January: 19–21.
Hammer, M. and Champy, J. (1993) *Re-engineering the Corporation: A Manifesto for Business Revolution*. London: Brealey.
Hayek, F. von (1949) [1945] 'The use of knowledge in society', in *Individualism and Economic Order*. London: Routledge & Kegan Paul, 1949.
Haywood, T. (1995) *Info-Rich – Info-Poor: Access and Exchange in the Global Information Society*, London: Bowker Saur.
Head, S. (1996) 'The new, ruthless economy', *New York Review of Books*, 29 February: 47–52.
Hirst, P. and Thompson, G. (1996) *Globalization in Question: The International Economy and the Possibilities of Governance*. Cambridge: Polity Press.
Hobsbawm, E. (1994a) 'Barbarism: a user's guide', *New Left Review*, 206 (July–August): 44–54.
Hobsbawm, E. (1994b) *Age of Extremes: The Short Twentieth Century*. London: Michael Joseph.
Hutton, W. (1994) 'Markets threaten life and soul of the party', *Guardian*, 4 January: 13.
Hutton, W. (1995) *The State We're In*. London: Vintage.
Information Infrastructure Task Force (1993) *The National Information Infrastructure: Agenda for Action*. 15 September. Washington, DC: NTIA NII Office. Internet address nis@ntia.doc.gov.
Kelly, K. (1995) [1994] *Out of Control: The New Biology of Machines*. London: Fourth Estate.
Kumar, K. (1995) *From Post-Industrial to Post-Modern Society*. Oxford: Basil Blackwell.
Labour Party (1995) *Information Superhighway*. London: Labour Party.
Lasch, C. (1985) [1984] *The Minimal Self: Psychic Survival in Troubled Times*. London: Picador.
Lasch, C. (1995) *The Revolt of the Elites and the Betrayal of Democracy*. New York: Norton.

Lash, S. and Urry, J. (1995) *Economies of Signs and Space*. London: Sage.

Luttwak, E. (1995) 'Turbo-charged capitalism and its consequences', *London Review of Books*, 2 November: 6–7.

Lyon, D. (1988) *The Information Society: Issues and Illusions*. Cambridge: Polity Press.

McKibbon, B. (1992) *The Age of Missing Information*, New York: Random House.

Mandelson, P. and Liddle, R. (1996) *The Blair Revolution: Can New Labour Deliver?* London Faber & Faber.

Mattelart, A. (1991) *Advertising International: The Privatisation of Public Space*, trans. Michael Chanan. London: Comedia.

Mitchell, W.J. (1995) *City of Bits: Space, Place and the Infobahn*. Cambridge, MA: MIT Press.

Negroponte, N. (1995) *Being Digital*. London: Hodder & Stoughton.

Newman, K.S. (1991) 'Uncertain seas: cultural turmoil and the domestic economy', in A. Wolfe (ed.), *America at Century's End*. Berkeley: University of California Press. pp. 112–30.

Owen, D. (1983) *Marxism Today*, March: 28.

Paxman, J. (1991) [1990] *Friends in High Places: Who Runs Britain?* Harmondsworth: Penguin.

Plant, S. (1993) 'Beyond the screens: film, cyberpunk and cyberfeminism', *Variant*, 14: 12–17.

Poster, M. (1995) *The Second Media Age*. Cambridge: Polity Press.

Postman, N. (1992) *Technopoly: The Surrender of Culture to Technology*. New York: Knopf.

Reeves, P. (1995) 'Gingrich sets up battlelines for the future', *Independent on Sunday*, 8 January: 1.

Reich, R.B. (1992) *The Work of Nations: Preparing Ourselves for 21st Century Capitalism*. New York: Vintage.

Rheingold, H. (1994) *The Virtual Community: Homesteading on the Electronic Frontier*. Reading, MA: Addison-Wesley.

Rheingold, H., Internet address http://www.well.com/user/hlr

Robins, K. (1995) 'Cyberspace and the world we live in', *Body and Society*, 1 (3–4): 135–55.

Robins, K. and Webster, F. (1988) 'Athens without slaves . . . or slaves without Athens? The neurosis of technology', *Science as Culture*, 3: 7–53.

Robins, K. and Webster, F. (1989) *The Technical Fix: Education, Computers and Industry*. London: Macmillan.

Rothschild, M.L. (1992) [1990] *Bionomics: The Inevitability of Capitalism*. London Futura.

Rowntree Foundation (1995) *Inquiry into Income and Wealth* (February). York: Joseph Rowntree Foundation.

Schiller, H.I. (1984) *Information and the Crisis Economy*. Norwood, NJ: Ablex.

Schiller, H.I. (1996) *Information Inequality: The Deepening Social Crisis in America*. New York: Routledge.

Scott, A.J. and Storper, M. (eds) (1986) *Production, Work, Territory: The Geographical Anatomy of Industrial Capitalism*. Boston, MA: Allen & Unwin.

Scott, J. (1991) *Who Rules Britain?* Cambridge: Polity Press.

Sennett, R. (1973) *The Uses of Disorder: Personal Identity and City Life*. Harmondsworth: Penguin.

Sklair, L. (1990) *Sociology of the Global System*. Hemel Hempstead: Harvester Wheatsheaf.

Slouka, M. (1996) [1995] *War of the Worlds: Cyberspace and the High-Tech Assault on Reality*. London: Abacus.

Stehr, N. (1994) *Knowledge Societies*. London: Sage.

Stock, G. (1993) *Metaman: The Merging of Humans and Machines into a Global Superorganism.* New York: Simon & Schuster.

Sullivan, W.M. (1995) 'Reinventing community: prospects for politics', in C. Crouch and D. Marquand (eds), *Reinventing Collective Action: From the Global to the Local.* Oxford: Basil Blackwell. pp. 20–32.

Talbott, S.L. (1995) *The Future Does Not Compute: Transcending the Machines in Our Midst.* Sebastopol, CA: O'Reilly & Associates.

Toffler, A. (1990) *Powershift: Knowledge, Wealth and Violence at the Edge of the 21st Century.* New York: Bantam Books.

Useem, M. (1984) *The Inner Circle: Large Corporations and the Rise of Business Political Activity in the US and UK.* New York: Oxford University Press.

Webster, F. (1995) *Theories of the Information Society.* London: Routledge.

Webster, F. (1996) 'The information age: what's the big idea?', *Renewal*, 4(1): 15–22.

Webster, F. and Robins, K. (1986) *Information Technology: A Luddite Analysis.* Norwood, NJ: Ablex.

Westergaard, J. (1995) *Who Gets What? The Hardening of Class Inequality in the Late Twentieth Century.* Cambridge: Polity Press.

Wintour, P. (1995) 'Labour looks abroad for ideas', *Guardian*, 6 October: 6.

Womack, J.P., Jones, D.T., and Roos, D. (1990) *The Machine That Changed the World: The Story of Lean Production.* New York: HarperCollins.

Woolgar, S. (1988) *Science, The Very Idea.* Chichester: Ellis Horwood.

Woolley, B. (1992) *Virtual Worlds: A Journey in Hype and Hyperreality.* Oxford: Basil Blackwell.

Wriston, W.B. (1992) *The Twilight of Sovereignty: How the Information Revolution is Transforming Our World.* New York: Scribner.

Zuckerman, H. (1988) 'The Sociology of Science', in N.J. Smelser (ed.), *Handbook of Sociology.* Newbury Park, CA: Sage. pp. 511–74.

PART 2

TEXTURES

4 Fishing with False Teeth: Women, Gender and the Internet

Simone Bergman and Liesbet van Zoonen

If anything can justifiably be called a 'technocity', it is the Digitale Stad (Digital City) of Amsterdam in the Netherlands. It was launched in 1994 on the initiative of Dutch hackers and the political cultural centre 'De Balie' in Amsterdam. Originally meant as a three-month experiment financed by the city council that would provide a new forum for political debate during council elections, it was so successful that the city kept its gates open after the elections. Its main goal has now become the introduction of Internet possibilities and facilities to a large public; the Digital City therefore provides e-mail services, discussion groups and an entrance to the Internet for free. Terminals have been put in public places to ensure that people who do not own a computer or a modem can gain access to the Digital City.

The interface of the Digital City is based on the features of ordinary cities. There are different theme-based squares, such as a central city square, a news and current affairs square, a cultural square and a gay square, which serve as meeting places for people interested in particular themes. Between the squares there are what would be houses in ordinary cities: homepages of inhabitants which contain information, personal stories, hyperlinks to other websites and many other things.

Amsterdam's Digital City has almost everything an ordinary city has, including the traffic jams at its entrance; in peak hours getting access can take up to 30 minutes. Nearly 15,000 people are in the Digital City every day: 9,000 of them are occasional visitors, the other 6,000 are the active part of the 20,000 officially registered inhabitants of the city. Because of the success of the Amsterdam Digital City, other digital cities have been created by local councils across the Netherlands. They mainly provide government, political and social information and lack the funding to set up the infrastructure for communication facilities.[1]

Some people doubt the success of the Digital City in Amsterdam. An extremely critical article in a local newspaper pointed to the loss of the city's public ideals: the city of Amsterdam often forgets to keep its information updated and the visitors and inhabitants do not seem to be very interested in political information or communication anyway. The public terminals were never adequately maintained and there is now only one left in the City Hall, maintained by the city council. The interface of the Digital City is terribly complicated, and mainly appropriate for high-speed hardware. According to the article, the figures on visitors and inhabitants obscure the fact that very few people actually participate in Digital City affairs; the majority of them use the e-mail facilities and the Internet entrance only. The news and discussion groups are said to be dominated by a small group of insiders who ridicule and marginalize potential newcomers. The article finally quoted the American magazine *Wired*, which seems to have claimed that the Digital City of Amsterdam is turning into a ghost town.[2]

The article, of course, evoked enormous controversy: active inhabitants dismissed much of the criticism as not specific to the Digital City but characteristic of Internet use in general, e.g. discussion groups being dominated by insiders. They claim that the Digital City is succeeding in introducing a substantial number of people to the virtual world of on-line communication and information; the number of visitors and inhabitants is exceeding all initial estimations.[3] The Amsterdam Digital City is in this respect more successful than its North American equivalents – so-called freenets – which offer on-line access to local government and organizations and provide a platform for digital communication.

Absent from this discussion is an evaluation of the demographic profile of the city users: a survey held by the Digital City has shown that they are predominantly male (90 per cent) and between 18 and 35 years old. The at that time female manager of the city (the 'mayor') ascribes this to the fact that the first places where you could get access to the Internet were located in technical universities, polytechnics and other workplaces which are dominated by men. That's why among the first group of Internet users there were so few women. She claims that among the non-traditional new users whom the Digital City, and the Internet in general, are attracting, the number of women is growing. It is not a fear of technology but a lack of appropriate software which keeps women at bay:

> It's not a fear of technology. Everybody has trouble working with the software you need for Internet: whether you're a woman or a man. The software is the most important barrier. It just has to work and then the Internet or the Digital City isn't man or woman unfriendly. (Marleen Stikker, mayor of the Amsterdam Digital City in Bullinga, 1995)

However, the user figures of the Digital City are too familiar to be so easily dismissed. It is not merely a matter of time, as the mayor seems to claim, before it will be a city for women and men. There are some specific features of the Digital City that hinder the easy access of women. The disappearance of public terminals is likely to affect the participation of women in particular. Previous research has shown that the presence of public terminals can reduce some of the barriers to female adoption of computer systems (Collins-Jarvis, 1993). The complex interface of the Digital City may also discourage women more than men because of their general tendency to doubt their competence to use computers. But even without such specific barriers, the Digital City would most likely be dominated by male users anyway given the more general phenomenon that the access to and use of new information and communication technology is structured along gender lines and other social divisions. In this chapter we will summarize and analyse the reasons for these patterns and discuss the various governmental and grassroots strategies that have been developed to overcome them. The first part of the chapter contains the more familiar debate on the exclusion of women from the new virtual worlds; the second part, however, will focus on women who actually do use new information and communication technologies. We will – as others have recently begun to do (Collins-Jarvis, 1993; Kaplan and Farrell, 1994) – claim that a theoretically and politically productive understanding of the relation between gender and technology cannot be built only on an analysis of the absence of women from technology. Instead a focus on what active female users of the Internet experience could be more productive in developing appropriate and effective counter-strategies and policy.

An Old Story for a New Phenomenon: Women and Communication Technologies

Irrespective of the particular time period they were introduced in, most new technologies have been accompanied by discourses of hope or gloom. The debates about moral decay as some warned, or social elevation as others would have it, supposedly resulting from the introduction of photography, cinema, telephone, electricity, radio, television, video, videotext and the new information technologies are remarkably similar (see Marvin, 1988). They evoke optimism about new and better futures for all of us or they prompt cynical and pessimistic responses that foresee a reinforcement of existing social wrongs at best, but consider the emergence of completely new kinds of injustice more likely. Debates about the information highway are similarly structured. Optimistic scenarios present the information highway as the solution to all contemporary social problems from pollution and traffic jams to unemployment and isolation. However, in pessimistic visions the information highway is presented as a First World, white and male domain which reconstructs existing international and social inequalities and introduces a new division between digital literates and illiterates: the information rich and the information poor.

There have been, of course, more subtle analyses of new technologies but they often get lost in the clamorous and more outspoken visions of bright or dark times to come. To the extent that women and feminists have taken part in these debates, they tend to side with the darker scenarios, claiming that the relation of women to technologies – new or old – is complicated and characterized by unease, tension, exclusion and exploitation. In most of these views, technology of whatever kind is seen as just another instrument and expression of gender dominance (for an overview see van Zoonen, 1992).

There is certainly a lot of historical support for such claims. To draw only a slightly exaggerated picture: the telephone was introduced as a medium for business and information, and the particular uses women found for the new medium were ridiculed and undermined (Rakow, 1988); the introduction of the radio in the domestic sphere transformed this women's domain at least temporarily from a place of conversation into a place of silence in reverence of a medium that needed a technician's (thus usually male) hand to operate it (Moores, 1988); many recent new technologies, in particular video, videotex and computer games, have heavily exploited the attractions of pornography to seduce a potential group of early adopters like young men; the fascination of home computers for men has again altered family life, with women having to compete with technology for the attention of the men in their families (Bergman, 1996).

Throughout the decades patterns of adaptation and use are remarkably similarly gendered for all of these communication technologies: business is the main mover of the introduction of NICTs, often producing the main early group of professional users (of telephone, videotex, computer, computer networks) who are, because of the nature of labour divisions, mainly men. But the information and communication technologies designed for leisure use also attract a group of so-called early adapters that are young, white and male. Even a family medium like television turns out to have been acquired for the benefit of husbands and sons to begin with (van Zoonen and Wieten, 1994).

Not surprisingly then, ordinary women and feminist scholars look upon the newest of information and communication technologies like telematics, digital networks, the Internet and other cyber inventions with reticence and suspicion, especially because applications such as the Internet seem to be as male dominated as their predecessors. The number of women active on the Internet is only a fraction of the number of men on-line. Although user statistics show a growing number of women using the Internet and percentages vary, they still account for only about 20 per cent of Internet use.[4]

Women and Technology: Explaining Absence

Several kinds of explanation for women's 'uncomfortable' relation with information technology are given in the literature. There is a popular commonsense explanation that refers to social-psychological factors to explain the absence of women: they are said to have less interest in technology, to lack technological skills and to perceive very few useful applications of information technologies. This is an assumption that is present in much governmental policy, as we shall see shortly. Feminists have strongly objected to such frameworks, both for blaming the 'victim', and for their denial of the gendered definition of technologies. If one defines technology in a more broadly encompassing way, women appear to be skilful and interested users as well as producers, for example in areas such as herbal medicine, agriculture and domestic technologies (Stanley, 1983; Faulkner and Arnold, 1985). Moreover, structural and cultural factors are considered to produce higher barriers for women than individual psychologies. Structural factors having to do with differences in education, income and social position provide very strong explanations indeed for women's typical relation to technology: on average women have lower incomes than men and therefore less to spend on computers, modems, software, on-line services and the other infrastructure necessary to get access to the information highway; on average they work in jobs and sectors in which they have less access to computer technology than many men; their position in the family deprives them of leisure time, etc. In other words, the fact that women use computer networks less than men do is partly explained by their structural position in society and the private sphere of the household.

As important as these structural factors are the cultural dimensions of technology, i.e. the way information and communication technologies acquire meaning as masculine domains and the way they function in the ongoing construction of gender identities. Feminist scholars in particular have enriched the debate on information and communication technologies by showing that not only are men the more frequent users, but men and masculine values also dominate the contexts of design, production, distribution and marketing, resulting in a very strong impression that these are technological artefacts made by men for men. Turkle (1988), for instance, argues that the gendered metaphors of control and submission that encompass computer technology hinder the development and recognition of computer skills that are based on cooperation and equality, supposedly more in line with what is currently thought of as feminine. Others have pointed to the context of production and design of computer technology, rooted in the military industrial complex (Wacjman, 1991). The popular figure of the hacker can also be seen as constructing computer technology as a symbol of inhumanity and alienation. These (masculine) meanings of technology thus function as a cultural barrier for women; information and communication technologies are seen as incompatible with women's activities and values. As Rogers (1983: 27) has shown, the acceptance and use of innovations is facilitated if they are seen as compatible with 'established patterns of behaviour for the members of a social system'. From the point of view of equal access and opportunities, women's rejection of computer technologies is a highly negative and problematic phenomenon. But looking at it from the point of view of how gender identities are being constructed and reconstructed in daily life, there is a more meaningful interpretation of women's reticence. 'Women use their rejection of computers to assert something about themselves as women: it is a way of saying that it is not appropriate to have a close relationship with a machine' (Turkle, 1988: 50). Thus, for women, rejecting information and communication technologies can be seen as a rational and positive choice, instead of as a sign of their exclusion or backwardness. In more theoretical terms, we can consider information and communication technologies as discursive practices in which the meanings generated in the production and reception process are negotiated, often resulting in the construction of a masculine domain (cf. van Zoonen, 1992, 1994; Bergman, 1996).

Women and Technology: Overcoming Absence

Several governmental policies have been developed to try to get women to use and enjoy new information and communication technologies in greater numbers; some of these policies have been part of a more general effort to engage women in technology. They vary from schooling projects, obviously aimed at removing some of the structural barriers, to public information campaigns trying to convince women of the necessity to join cybersociety. In such campaigns the implicit assumption is that women make the wrong choices in education or in the job market. A campaign to convince girls to choose mathematics and other similar subjects as exam options in secondary school carried as its slogan 'Kies Exact' (Choose Beta-Subjects), thus presenting the gendered nature of these subjects as a matter of individual choice, neglecting the structural and cultural factors that produced this gendering in the first place and ignoring the problems girls encounter when opting for masculine domains and the benefits they get from more 'traditional' options. Another campaign trying to persuade women to choose a traditionally male field of work wasn't accompanied by a similar campaign trying to convince boys to take on more domestic work or to engage in traditional female labour areas. Thus, most public information campaigns construct the problem of the relation between gender and technology as the result of individual psychologies of women who have the 'wrong' preferences and make the 'wrong' choices.

Grassroots campaigns, set up by feminists who have mastered and enjoyed new information and communication technologies, have taken another road: they typically point at the instrumental and/or pleasurable uses of new information and communication technologies and offer practical guidance to get access to them. An example is the *Woman's Guide to the Internet* which was published in 1995. The guide, in brilliant pink, provides an introduction to the Internet. It is not so much a manual on how to use the Internet as an explanation of what the Internet is, its history, the kind of information on it and how you can communicate through it. Lastly, the implications of the Internet as a precursor of the information superhighway are discussed.

The author, Marianne van den Boomen, claims that whereas the guide is aimed at women in particular, it can be useful to everyone. Other books in this genre are written for people who want to start using the Internet immediately. The author notes the fact that these manuals are mostly read by men:

> The people who read the Internet manuals are for some reason more often male than female. Apparently women are less eager to use new technical gadgets. But the Internet isn't a gadget for specialists and hobbyists anymore: it is finding its way into our daily lives. And in their daily lives more and more women are dealing with the Internet. In their work, in their education and via friends. (van den Boomen, 1995: 10)

Van den Boomen asserts that the fact that women are hesitant either to learn about the Internet or to go on-line is the result of the reticence women show towards technology in general, and towards computing and computer networks in particular. Women first want to know what you can use the Internet for, before they will try it out. That is why, according to van den Boomen, this guide offers an overview of the practicalities of the Internet, explains why it's useful and gives an impression of its atmosphere. In the guide she shows why the Internet is of practical use in various ways, like lobbying, networking, exchanging information and coordinating actions, all instruments which are very useful, for example for the women's movement. Next she shows how the Internet can be fun: browsing in a bookshop and sending someone you haven't seen in a long time a postcard. The book thus can be seen as an attempt to reconstruct the Internet as a technological practice that is compatible with women's activities, needs and values.

The book caused an interesting reaction among some of the inhabitants of the Digital City who gather in one of the 40 discussion groups, dds.femail, for and about women, but open to men also. These were established when a number of women were sexually harassed via e-mail and wanted to share their experience. Now there are many more topics debated in dds.femail, for example emancipation, psychological differences between men and women, feminism and birth control. The debate on the *Woman's Guide to the Internet* centred around the question of whether women need a special guide. Isn't the unspoken assumption that women are too dumb to find out themselves? Men in particular objected to making a gender difference among Internet users, either by denying the problem or by ridiculing women who do not succeed in going on-line. In the box are some fragments from an electronic response the author of the book gave to some of the reactions that were posted on the list.

Box 1

Reaction to the book:
Does the internet work differently if you're a woman? Is it all of a sudden faster?

Author's reply:
If it only would! Then there would be many more women on the Net than the 15 to 20 percent of the Internet users at the moment. Because most women have got something else to do (usually a thousand things at a time), instead of endlessly trying to learn about that damned Net. Women are not afraid of technology or the Internet; they are afraid to waste their valuable time. They first want to know whether it's useful before they start to touch the buttons. And that's what my book offers. For the people that first want to know what they're starting at.

Reaction to the book:
So if I understand correctly women need an extra booklet in order to understand the Internet?

Author's reply:
Yes. That is, in order to understand *rapidly* everything about the Internet and in order to get rid of some common off-line prejudice about the Internet. Because they don't want to waste any time. Just like the way women learn Word Perfect when they need it for their work or education, in one afternoon. While many men have been trying to learn it by trial and error which takes years.

Reaction to the book:
Yeah, well that's how it goes. I personally have been looking for a book called 'Fishing for people with false teeth' and I'm especially eager to find 'Feminist chicken breeding for the left-handed'.

Author's reply:
If you look hard enough you can probably find it on the Internet. Keep trying!

Reaction to the book:
I also think women shouldn't be afraid of computers. Computers are useful and you can learn from them. And there are lots of computer programs which are quite easy, like Windows, so nothing can go wrong.

Author's reply:
I don't assume women are afraid of computers. And neither do I think everything should be made as simple as possible for women. But indeed computers are useful and Internet is a fantastic invention.

Woman:
As a matter of fact I have a lot of friends who don't want to have anything to do with it. But if I tell them about Internet, they always would like to try.

Author:
Yes, I also know women who don't want to have anything to do with it, or can't imagine why it's fun. But they are not afraid of computers. And it doesn't have to be simpler especially for women. They are no more afraid of technology or more stupid than men are.

The few reactions quoted here show a whole spectrum of denial and negation strategies in the public debate on gender and new information technologies: there is an assumption that technology itself is neutral ('Is it all of a sudden faster?'); there is the idea that women have a different relation to computers than men do ('so if I understand correctly women need an extra booklet in order to understand the Internet?); and there is of course, ridicule ('Fishing for people with false teeth').

Let us stop here for a moment to consider how the debate about women and information technologies is framed. We wrote: the number of women is only a fraction of the number of men; women account for only 20 per cent of Internet use. The absence of women is caused by structural and cultural barriers that produce a psychologically valid choice for women not to take part in information technologies. Governmental policy to overcome the absence of women in information technology usually tries to persuade women and girls to become active users of technology, assuming that it is a mere matter of choice. Grassroots feminist strategies highlight the instrumental and/or pleasurable uses of information technologies and offer practical guidance to get access to them.

Both options have in common that they take the absence of women as their point of departure. Policy and strategies are thus tuned towards the absence of women and the 'problems' produced by it. It seems that at least two groups are ignored here: the astonishingly high number of men who do not use new information technologies and the women who do use ICTs without many problems. On the Internet, for instance, 20 per cent is a figure that is probably on the low side because of the ample opportunities to play with gender identities on the Net (e.g. Rogerat, 1992). Also, there is much research that shows that in professional contexts there is very little difference between women and men as far as the use of ICTs is concerned. By ignoring such figures, academic debate, research and policy on gender and information technologies contribute to their construction as a masculine domain, enlarging the distance between women and technology further. In addition, it does seem rather strange to build theory and policy mainly on an absence rather than a presence. Therefore, in the second part of our chapter we shall focus on one of the two ignored groups in the debate on gender and NICTs: women who do use NICTs quite unproblematically (leaving the non-technological male to other researchers for now).

We have started interviewing three women who are on-line. Our results are fairly provisional and are aimed at providing some new ideas for research and policy rather than providing definitive data. We will begin by describing how these three women came to use computers and computer networks. Next we will describe for what purpose they use computer networks and determine the meaning of computer networks for these women. Finally, we will discuss their experiences of being a woman on-line.

Women On-line

Marjorie is a 35-year-old lesbian woman who lives with her partner in the east of the Netherlands. She has a doctorate degree in social linguistics and works at a research institute. She uses the computer in her work and got acquainted with the Internet via a colleague. She has a computer and a modem at home. Julie is 30 years old and lives with her husband in Amsterdam. She is a graphic designer and started to study computers in a more 'serious' way after she had done some courses in desk-top publishing for her work. Recently she bought a modem, subscribed to a provider and started to use the Internet. Monique is a 32-year-old woman who lives alone in the Hague. She runs her own bulletin board system (BBS) at home. She used to work in word processing but at the moment she works at a computer helpdesk. We met Marjorie at a conference on new media. Monique and Julie responded to a message we posted in dds.femail asking women who are on line to respond.

Going On-line

Marjorie had been using the computer for word processing when she got in touch with computer networks at her work a couple of years ago. As a researcher at a research institute she had access to the Internet via a UNIX computer. She had heard about the Internet and was fascinated by the idea of communicating with people all over the world. With the help of a colleague she took her first steps on the Internet. He gave her some UNIX commands and then let her find out by herself. She started trying and although initially – as she said – she didn't know anything, she soon got the hang of it. She finds it reassuring that she can always ask her colleague for help when she has a problem with her equipment. But as she says: 'He's very strict. He will first let me try it by myself, and only if I don't succeed will he come and help me.'

Initially she used the Internet mainly to mail colleagues, but gradually she started to use the Internet not for work but for personal reasons. While she only scans mailing lists related to her work for useful information and never contributes or poses questions, she is an active member of other lists varying from African music to writing. She is even a celebrity on a lesbian mailing list because she posts so often in this group. With some of the women on this list, who live abroad, she has a frequent and meaningful e-mail friendship. Because she has recently started to work a lot at home, she now has Internet access at home. Her modem is not fast enough to 'surf on the Net' – which also might lead to a huge telephone bill – so she makes very limited use of the World Wide Web. She mainly uses the Internet for communication.

This plays an important role in her daily life. Not only does she spend several hours a day writing and replying messages, but this contact has a great impact on her:

The impact of these messages can be enormous. You can see something on the television, but it won't have the same effect on me as when somebody writes me. I mean, last week I received a message from Israel, saying that just around the corner another bomb exploded. A message like that, written by people who live on that very spot and who write about their own personal emotions, is so different from a news broadcast about an explosion.

As a graphic designer Julie is very experienced with computers. She did a course to learn how to work with different desk-top publishing software. Although she realized at work that she is good at working with computers, she wanted to know how they actually operate. At the same time she realized that her job wasn't satisfying enough. When she noticed that people who studied computer science had better job opportunities, she decided to go to the Open University to study computer science. She had read a lot about the Internet and decided that because of her study she had to experience it for herself. She bought a modem and the appropriate software and installed everything at her computer at home. She describes her attitude towards the Internet as follows:

> I think it's something I just have to know: it's inevitable. They're a necessary evil. It's like a washing machine: they're no fun, but quite handy. I realized I couldn't stay behind. At first I didn't really know what the Internet was about. But now that I'm actually on it I really see the possibilities and I really like it! Of course it does cost a lot of money, but I've saved some money so that I could experience it myself.

Although for Julie it was almost inevitable that she would go on-line, because she studies computer science and therefore just 'had to know' about the Internet from experience, it was something she decided herself and did by herself. There were no other people around who inspired or helped her, as in Marjorie's case. Now Julie spends a couple of hours behind her computer every day. She finds the Internet very useful for instance for her studies, for downloading computer programs and to keep in touch with her teacher at the university. At her university there are several bulletin board systems which she also uses. And although at first she wasn't interested in discussion groups, now she is even thinking about starting her own because she likes, as she says herself, to chat about all sorts of things and to provoke reactions.

For Monique computers have been a hobby from an early age. Initially she had a home computer which in time she replaced by a personal computer. Her interest in computers and software grew over the years. When she was working in word processing she became interested in administrative computer programs which she downloaded from different bulletin board systems. At home she also spent several hours a night downloading software. Because of the telephone costs this was getting very expensive and she decided to start her own BBS so she could download software by herself. On her BBS she offers e-mail and discussion groups, but also games and other software. She herself uses Internet mainly for communication, either person to person or in newsgroups. She can spend hours replying to messages from her BBS users, or if she wants to discuss something on television she simply posts a message to a newsgroup waiting for a reply. She finds information on the World Wide Web less interesting, although she sometimes does invest time in reading something. She spends at least two hours per night on weekdays behind her computer. As soon as she comes home from work she checks who called her BBS. After supper she checks whether any new software programs have arrived and tries them out. In the meantime she has turned her hobby into her work: she did a course on system management and is now working at a helpdesk.

These women all had their own valid reasons to go on-line. For Monique it was the result of a growing hobby; Julie wanted to keep up with the developments because she started to study this field and Marjorie went on-line because she was fascinated by the idea of communicating with people all over the world. All of them especially value computer networks for communication. Let's have a closer look at why they are so enthusiastic about electronic communication.

Electronic Communities

Monique usually communicates with women who call her BBS. These conversations are usually about personal matters, like sexuality and relationships. She states that making contacts and communicating via computer networks is easier. As she says: 'In real life I would not go up to a woman and say do you fancy a talk?' It's a thrill for her to post a message and wait for a reply. She likes the idea of having conversations with people she has never seen, because of the anonymity. For Monique communicating via e-mail is a relaxed way of spending her time. She reads and writes her messages off-line and can take as much time as she wants for it. She puts a lot of effort in putting her thoughts into words. Sometimes she spends several hours thinking about how to write a particular message.

Marjorie shows the same care in the writing of her messages. She compares writing via e-mail, as she does with several good friends, to the culture of writing letters: 'It's nothing new, really. It's just like letter writing in the eighteenth century, only much faster. Usually you try to write with very much precision: very balanced and well considered. I think that's the same culture.'

As a result of her participation in one particular discussion group Marjorie has friends all over the world without actually having to visit them, something which is quite convenient for her since she is afraid to fly. Marjorie finds it an ideal way of communicating: while she is hesitant to call someone, she will approach people by e-mail without a problem. She also uses e-mail to keep in touch with her brother who actually lives nearby. E-mail and the lesbian discussion list have become an important part of

Marjorie's daily life. This communication is not only very frequent, it is also very intense. As she puts it:

> It's much closer to you and more real. Like discussions, even if you only read them, you don't just happen to overhear them in a pub or something: it's right there under your nose! You can either participate, or just read them. But either way, you will continue thinking about it. Your mind doesn't stop. In this direct way you get a lot of information you wouldn't have normally.

For Julie, commmunication wasn't what triggered her to go on-line. Mainly she was interested in what the Internet was all about. At first she felt uncomfortable about the idea of communicating with people she had never met: 'It's a new way of thinking and I also realized I was talking with people who were complete strangers to me! That was quite strange: receiving mail from people you don't know.' Now she finds communicating via computer networks an accessible way of keeping in touch without actually talking to someone, for instance over the telephone. Compared to the telephone she states that you can be more precise in what and how you want to say something: it is a more relaxing way of communicating. Julie thinks computer networks can be quite useful for networking and is planning to do this. From her own experience she noticed that this is actually a very good way to make contacts. During an airline flight she met a man. They talked and he offered her a job. When she returned home she sent him a message:

> I would never have called that man. Yeah, I can be loud-mouthed, but sometimes I don't know what to say. E-mail is accessible and less scary.

It is obvious that Marjorie, Julie and Monique have found a meaningful way to use the Internet. Let us have a look at their first steps: were they hindered by the barriers that are commonly believed to keep women at a distance from computing and computer networks?

Taking the First Steps On-line

Neither Marjorie, nor Monique, nor Julie reports having had major problems in going on-line. They were all used to working with computers, which made the step to the Internet easier. Marjorie was introduced to the Internet by a colleague. Both Monique and Julie had to find it all out by themselves. This however didn't cause a lot of problems for them. Almost to Julie's surprise the installation of her modem and software went very smoothly:

> Well at first something went wrong but when I found out what the problem was it all went fine. Everything is working perfectly! I've heard everyone has lots of trouble getting everything to work properly, but I haven't experienced any problems!

Monique tried to find out everything by herself. As she indicates, her knowledge of the computer has grown over the years. Instead of taking her computer back to the shop when something went wrong, she now fixes everything herself:

> Well, eventually after you're more experienced, you try to find it out by yourself if something goes wrong. In the beginning I didn't dream of unscrewing my computer: I used to bring it back to the store. Now, the outside of my computer is underneath the table and I can look right into the computer. In the beginning that was a bit scary. But if something goes wrong, well no problem, you put everything back in its place and start all over again.

Julie, Monique and Marjorie don't seem to make a big fuss about difficulties they experienced in going on-line. The fact that they sometimes don't know all the details on how and what to do, didn't prevent them from taking the step to computer networks. They just try to find out by themselves. And as Marjorie puts it:

> There really isn't that much to know. Once everything is installed, you don't have to know how to do anything. It's the same with everything else you have to learn: once you do something often you'll learn it. It's like driving a car: you don't have to know how a car works to drive it. That's what mechanics are for!

These women seem to do fine on-line. Could we then conclude that the fact that they are women doesn't make a difference in going and being on-line? Doesn't gender play a role? Let's have a closer look at how these women feel about their position on the Net.

Being a Woman On-line

Julie definitely has the feeling that there aren't many women on the Internet, as shown from the fact that she usually receives mail from men. As she says: 'It really feels like a man's world.' This however isn't a problem for her, because she is used to being a woman in a man's world. In her work she is also one of the few women and although she is aware of the fact that she is different in this world, it isn't a problem because she can deal with that very well.

Monique, being not only a woman active on the Net but also one of the very few female system operators of a BBS, is also very aware of her 'special' position. She recalls reactions from other system operators when she first started her own BBS:

> You get reactions like 'Oh, can you also run a BBS?' I said 'Sure! Why not?' But you always get those kind of remarks. Like they think 'Oh, you're a woman and you're not supposed to know anything about computers'. I say to them: 'Why are only you men allowed to do this?' I just like to do it and think it's fun: it's my hobby! It's bullshit that only men should be on the Internet.

Because of the anonymity she doesn't bother too much about possible harassment:

> I use my own name. But they don't know anything: they know I'm a woman and they know I'm called Monique. That's all. In that respect I'm really down to earth: I'm a woman and I'm on the Internet and I'm proud of it. Everybody may know that! And of course there are more men: but that gives you a chance to become unique and that's nice too!

Marjorie also acknowledges that women in general are a minority on the Net, and lesbian women even more so. This means, for example, that you meet the same women on different discussion lists. The fact that few women have access to computer networking does not mean that only a few women share the information and communication. Marjorie advocates the idea that the women who are on-line should be intermediaries for the women who are not on-line (yet). This way the information on discussion lists has a far larger distribution than the women on-line only: 'The women that do have access to the Net and that have political ideas, usually women from minorities like lesbians, have a enormous amount of women backing them, who don't have access to the Internet.'

This way the Internet can be very useful for feminist and gay politics, as Marjorie's activities demonstrate. Organizations active in this field should have their own homepage in order to distribute information widely. Marjorie notices that discussion groups pick up political matters or other issues that are related to the gay scene and make them known more publicly or even incite action:

> In that discussion list everything which is relevant to the gay community is discussed. Even demonstrations are announced. Either on the Internet or by telephone. The news spreads like wildfire. And the impact is really worldwide and even on the national level it works.

None of the women we interviewed encountered the barriers which are said to prevent women from participating on computer networks. They do experience the Internet as being dominated by men, but they employ their own tactics for coping with their isolation and turning it into a perceived advantage. They obviously do not submit to male dominance, but have found their own way of using the Internet: they have reconstructed so to speak the masculine meaning of the Internet by using it in a way that suits their interests and needs. Their use is not in conflict with their daily life, but has become an integral part of it.

Conclusion

The experiences of these three women shed another light on the social-psychological, structural and cultural barriers women are said to face when trying to go on-line. Marjorie, Monique and Julie are all in their thirties, living alone or with a partner without children. They are employed part time or full time. Being single or married without children, their home environment is different from that of the traditional housewife and mother assumed in most studies on women and information technology. They have more time and money to spend on information technology. Also all three are in work or educational situations which give them easy access to computers and computer networks. Structural factors do not seem to produce barriers for them.

Looking at the way they use Internet facilities, they have found activities which are personally and politically relevant to them and can hardly be said to be part of a masculine culture. Marjorie uses computer networks for feminist and gay politics, while Monique has found a way to let others benefit from her hobby by running her own BBS. And all of them enjoy e-mailing with people all over the world.

The communication facilities the Internet provides obviously are compatible with the needs and values of these three women, making their acceptance and use of the medium unproblematic. But their fascination with the communication facilities of the Internet has broader relevance too: the communication facilities of the Internet can be seen as the virtual translation of more or less traditional feminine concerns of personal contact, sharing and creating community. In effect, Marjorie compares her e-mailing with the female tradition of writing letters that dates back to the eighteenth century. But Internet communication also appears to lead the possibilities of communication in directions that many women (and men for that matter) have found problematic. The sheer informality, anonymity and virtuality of Internet communication obviously helps women to make and follow up on contacts with strangers.

Marjorie's, Monique's and Julie's personally meaningful uses have turned the Internet into a psychologically rewarding experience for them that isn't at odds with the rest of their lives, or their gender identity. Their experience shows that there is a lively 'feminine' culture on the Internet that needs to be revealed in more detail in order to 'demasculinize' the Internet.

This demasculinization could be part of wider measures that employ analyses of women's absence as well as their presence on the Internet. The results of our study suggest a rethinking of such measures in several directions. To begin with, it seems smarter to direct policy and strategies at women who are close to information technology anyway than to try to convince women for whom the structural and cultural distance from information and communication technologies is bigger. However, if one does want to develop new overall policies other than the ones already in place, public access terminals should be an integral part of them, and these should be placed in locations where many women and other unlikely Internet users gather. For those completely uninitiated women, but also for women who are familiar with computers but are reluctant about using the Internet, the software needed for the Internet and the interfaces used by the service providers should be as user-friendly and as adaptable as possible. This, of course, would be beneficial to a variety of people who are far removed from the experienced young white males at present dominating computer culture. At present, none of the more than 500 service providers in Europe have developed special policies for their women clients nor for other unlikely Internet user groups, and they are not likely to do so without market or governmental pressure.[5]

Lastly, and probably most importantly for new kinds of policy, is the recognition and publication of the presence of women on the Internet and the specific uses they make of it. It would be part of a new mythology to claim that these are specifically or essentially feminine activities. However, women's pleasures in the Internet could be instrumental in showing the enormous diversity of Internet activities and the potential benefits for our multicultural societies. In such a context, emphasizing the compatibility of the Internet with a historically specific set of women's needs, activities and values will be more productive than to keep pointing at the masculine character of the Net, thereby maintaining the mythology of technology as predominantly male.

Notes

1 Other Digital Cities are Eindhoven, Leiden, the Hague, Rotterdam and Utrecht. These are the major cities in the Netherlands. Recently other cities have also been developing their digital equivalent.

2 *Het Parool*, 18 November 1995.

3 *Het Parool*, 25 November 1995.

4 In various surveys on Internet use the percentage of female users ranges from 15 to 25 per cent. See also the following demographics: City.Net *http://www.city.net* /cnx/survey_fall95.html), GVU's WWW user surveys, Nielssen Media Research (http://www.nielssen.media.com/whatsnew/margs.htm) and Internet Connection (http://www.internetconnection.com/dmgrph.htm).

5 Personal communication with a coordinator of the activities of European service providers.

References

Bergman, S. (1996) 'Communication technologies in the household: the gendering of artefacts and practices'. Paper presented at the Granite/COST A4 seminar 'The shaping of gender and information technology in daily life', Amsterdam.

Bullinga, M. (1995) 'Digitale steden en dorpen. Nederland staat in de steigers' (Digital cities and villages. The Netherlands under construction), *Tele PC*, August.

Collins-Jarvis, L. (1993) 'Gender representation in an electronic city hall: female adoption of Santa Monica's PEN system', *Journal of Broadcasting and Electronic Media*, Winter.

Faulkner, W. and Arnold, E. (1985) *Smothered by Invention: Technology and Women's Lives*. London and Sydney: Pluto Press.

Kaplan, N. and Farrell, E. (1994) 'Weavers of webs: a portrait of young women on the Net', *Electronic Journal on Virtual Culture*, 2(3) (ftp:\\byrd.mu.wcnet.edu).

Marvin, C. (1988) *When the Old Technologies Were New*. New York: Oxford University Press.
Moores, S. (1988) 'The box on the dresser: memories of early radio and everyday life', *Media, Culture and Society*, 10: 23–40.
Rakow, L.F. (1988) 'Gendered technology, gendered practice', *Critical Studies in Mass Communication*, 5: 57–70.
Rogerat, C. (1992) 'The case of Elletel', *Media, Culture and Society*, 14: 73–88.
Rogers, E.M. (1983) *Diffusion of Innovations*. New York: Free Press.
Stanley, (1983) 'Women hold up two-thirds of the sky. Notes for a revised history of technology', in J. Rothschild (ed.), *Machina ex Dea. Feminist Perspectives on Technology*. New York: Pergamon Press. pp. 82–98.
Turkle, S. (1988) 'Computational reticence: why women fear the intimate machine', in C. Kramarae (ed.), *Technology and Women's Voices. Keeping in Touch*. London: Routledge. pp. 41–61.
van den Boomen, M. (1995) *Internet ABC voor vrouwen. Een inleiding voor datadames en modemmeiden* (The Internet Guide for Women). Amsterdam: Instituut voor Politiek en Publiek.
van Zoonen, L. (1992) 'Feminist theory and information technology', *Media, Culture and Society*, 14: 9–29.
van Zoonen, L. (1994) *Feminist Media Studies*. London: Sage.
van Zoonen, L. and Wieten, J. (1994) 'It wasn't exactly a miracle. The arrival of television in Dutch family life', *Media, Culture and Society*, 16: 641–59.
Wajcman, J. (1991) *Feminism Confronts Technology*. Cambridge: Polity Press.

5 The Ideal City and the Virtual Hive: Modernism and Emergent Order in Computer Culture

Julian Stallabrass

'Observers of media art,' wrote Herbert W. Franke, one of its most venerable practitioners, 'notice that a new turning point has been reached. Perhaps this will lead to a decisive breakthrough. Art critics and philosophers are proclaiming the dawning of a new epoch, a "second modernism" characterised by the application of the new media' (Franke, 1996: 253). And, indeed, it is true that computer art now seems ubiquitous in exhibitions of contemporary art, and that over the last few years the literature dealing with computer culture has grown exponentially, and that it is full of proclamations.

What can we learn about the ideology of network and computer culture from examining the way it looks? The visual element in computing has assumed ever greater prominence, not only in business software but in the games which play a leading role in driving the technology standards ever upward. These looks may, of course, be taken as symptoms, but they are more than that. Given the concentration on the visual in the culture generally, and the increasing precipitation of culture at the level of the surface, the visual aspect of computer culture is no epiphenomenon. This chapter will look briefly at the visual aspect of contemporary interfaces, but to examine tendencies and future developments it will also examine the history of computer art, and the virtual environments constructed in current leisure software. Both are primarily focused on the visual, and both have developed styles and habits which affect computer culture as a whole.

Walter Benjamin made a long study of the nineteenth-century arcades of Paris, by this time housing a run-down collection of shops, from which he elicited visions of a transparent and transcendent architecture; despite their ostensible use, he thought, we could read from the structures of these buildings the utopian, universal moment, the broken promise, of bourgeois ideology.[1] It was a way of looking back to also look forward. In the same way, we may look to the Paris of Benjamin's time, to modernist visions of the 1920s for similar, and more explicitly utopian, material. Although these modern visions, from Le Corbusier onwards, have been subject to systematic and often effective attacks in other areas of fine art and architecture, they retain a remarkable affinity with much current computer art.

If we think first of Le Corbusier here, it is because he is the originary figure of utopian modernist architecture. He made of modernism's more rationalistic precepts a systematic and universalizing system. In his city plans, zoned and functionally differentiated areas, standard units, and utterly straight roads, he created an order which was simultaneously social and aesthetic. These plans of the early and mid-1920s, such as 'Une Ville contemporaine' (1922), were a comprehensive attempt to solve material urban problems and to achieve social harmony and spiritual well-being.[2]

Although they were out on a technophile limb, Le Corbusier's recommendations were part of a wider conservative project current in France after the First World War which sought to bring anarchic, bohemian modernism into line, and which, on Jean Cocteau's agenda, sought to develop a 'degree-zero' language of form, a bureaucratic manner, so self-effacing as to escape the very description of 'style', allowing the message to flow directly forth.[3] Le Corbusier thought of his architecture as styleless, a flowering of rational thought about human needs and urban problems solved once and for all, with the additional leavening of a quantity of architectural genius. The simple primary forms of architecture or painting would speak universally and directly to the human organism.

Those of Le Corbusier's drawings which show his city plans from a distance resemble microprocessors – and this similarity is not fortuitous since everything was done to ensure that traffic moved unimpeded at the highest possible rate to and from the central administrative core to the zoned, functionally differentiated elements surrounding it. Compact residential units or office towers were to be set in parkland, a contrast of opposites, zero and one. There was no room for (analogue) suburbs. As in the processor, the raw measure of efficiency was speed. The materialization of this efficiency would produce beauty.

But it was not enough that Le Corbusier's early designs were rational expressions of what he took to be human needs. Given the cultural and economic obstacles to the immediate implementation of his plans, they also had to act as propaganda. In some of Le Corbusier's early villas, conventional frame and brick-infill construction (used because it was cheaper for a single building than concrete) was covered with a unitary layer of plaster, to suggest that the wall was composed of a single surface. So, as well as being rational, his buildings had to look rational, even at the expense of a little dishonesty.

Such a technique, in which the aesthetic retires, leaving in its stead the very image of rationality, is a familiar modernist trick; and is seen still in many aspects of our contemporary culture (car design, for instance, which makes a point of looking rational, as do uniformly bureaucratic-looking system boxes and monitors), above all in software interfaces. The image of rationality condenses on the screen: Windows, the ultimate triumph of form over economy, poses as a rational system. Think of its particular features which have become so familiar that we tend to take them for granted; and how familiar it is all designed to be – the files and folders, those little thumbnail sketches which so appropriately bear the name 'icons' (and these little pictures are often of familiar objects), the sculpted 3-D buttons, the pop-up notices, decked out with instantly recognizable, if not comprehensible, symbols warning of hazards or admonishing the user's mistakes.

The extraordinary advance in computer technology has largely served to bring to the 'desk-top' ever more sophisticated interfaces which throw an analogue cloak over digital operations in an attempt to convince the user that handling a computer is some simple craft skill, largely dependent on manual dexterity. Endless concessions are made to computer 'dummies', to reassure them before this otherwise unyielding, even unnerving, device.[4] In an extraordinarily patronizing display, the act of copying a file in Windows 95 brings up an animation showing pieces of paper flying from folder to folder. The constant cheerfulness of the interface works against deep fears of inadequacy that may, as we shall see, be justified.

The images of rationality and familiarity, then, are thrown together. The latter is supposed to mitigate the alienating effect of the former – not just 'computer', says Windows 95, but 'my computer' – but the image of rationality can never be disposed of because it is at the core of the enterprise. It is also much needed as a mask. For, despite all the effort expended on it and the computing power thrown at it, the software industry's standard programs are still fearsomely complex and quirky beasts: why, asks Clifford Stoll (thinking of Microsoft), does he need to have a 1,000-page manual to find out how to write a letter to his friend Gloria? (Stoll, 1995: 66).

But surely, despite certain similarities – rationality taking the form of an aesthetic, and an aesthetic serving to mask the absence of rationality – to talk of the Le Corbusier of the 1920s is to talk of another age. His totalizing schemes; his view of the aesthetic as an autonomous but complementary supplement to the rational which can, however, be objectively judged; his faith in the power of human reason and technology to solve the most pressing and difficult problems – all these views must have long since been buried by the voluminous critiques of postmodernism. But not, if you care to look, in computer art. There, highly traditional concepts of beauty, long called into question in other areas, still abound, and sophisticated technical means are frequently married to naive aesthetic projects. Computer art has often been based upon idealist aesthetics, using explicitly Platonist models relating number and beauty. Its theorists frequently claim that the new technology has made possible objective progress in aesthetics.[5] Around computer art there is much talk which Le Corbusier would have found highly familiar, of aesthetic laws and their discovery, of mathematically generated or governed art. (This is not to say that such speculations are necessarily wrong-headed, but they are certainly modernist.)

Computer art is also dominated by an old-fashioned impulse to naturalism, to giving an immersive experience in 16.7 million colours at whatever pixel resolution (it has been calculated) necessary to make 'reality' appear at

24 frames per second. The measure of this naturalism is interesting, especially given the grounds on which the notion has been so frequently criticized: it is the physiological capabilities of humans. The ambition is generally to make a seamless, unitary environment in which the participant is inescapably immersed. As in high modernism, a technophile utopianism is abundant, one which assumes that art, and even what it means to be human, will be radically recast through the medium of technology – and this is a view naturally much encouraged by the computer industry.

What computer culture offers is the possibility of making an image of the ideal appear real. So the construction of paradises made for exploration is a strong urge. Common are dreams of immersion and free flight, of soaring above unspoilt natural wilderness and sublime modern environments. Urban or pastoral, both visions are modernist, since they base their utopias upon a technical fix which carries us either to the Heavenly City or the Garden of Eden (Benedikt, 1991). They are as complementary as the relation of parkland and high-density tower blocks in Le Corbusier's schemes, being interdependent, inverse images of one another. In these digital scenarios, the promise of aesthetic harmony and an implied social calm, built on the foundations of a thoroughly linear system, produces a strong sense of *déjà vu*.

The same naivety is found in much network culture; obviously a good deal of it is devoted to deliberately ridiculous trivia, or porn, or gaming, but to look away from the material to the unifying interface in which it is framed, the visual expression in the Windows or Mac systems is of a benign technocratic modernism; of ease and comfort in exploration, along with allusions to space travel. And it is travelling, as much as arriving, which is important: as for Le Corbusier, for whom cruising along the motorways of his city at the incredible speed of 60 miles per hour would allow the traveller to take in the order of the entire urban composition and to be in harmony with it, so in satisfactory Net exploration (rare, it must be said, in this country) it is the sensation of speed, of real matched to virtual geography that is often important (Le Corbusier, 1971: 177). Also, of course, that old modernist paradigm – discovery.

The powerful element of unashamed modernism in network and computer culture is surprising, if only because of the frequent theoretical claims that this is the pre-eminent arena of postmodernism. For instance, in writing of virtual 'rapes' on a MUD (Multi-User Domain), one author recently warned against thinking of cyberspace as a utopia, for 'the wounds of modernity are borne with us when we enter this new arena' (Poster, 1995: 86). To say that they are 'wounds carried with us' is to think that the wounding itself is over and done with, and that the injuries will eventually heal in the balming virtual environment. But computer culture is not contingently modernist: it is founded on a technology which is a living embodiment of the modernist, positivist dream of directed evolution and apparently limitless progress. If this outlook is mapped back on to the poor, static or evolutionarily slow human users of this giddily advancing technology, and allied to the highly 'reductive' science of genetics so that people begin to talk seriously about the 'post-human' (as they once did of the 'new man'), that, too, is not an accidental matter.

Yet computer culture now has not simply readopted modernism as if postmodernism had never existed. Modernism has been revived but has also been altered in the process. To understand this, we need to look briefly at the dystopian vision in computer culture. Apocalyptic elements are so prominent that it has become something of a cliché to think of computer culture as held between the poles of heaven and hell, utopia and dystopia. I want to highlight this issue by looking at the work of William Latham, one of the most prominent and successful British computer artists. Latham held a fellowship at IBM during which he collaborated with their programmers to develop his particular style and manner of working. His pictures and videos are known beyond the ranks of technophile high-art enthusiasts – through an album cover for Shamen, and the subsequent adoption of his work by techno enthusiasts, and also because Latham is assiduous in courting the mass media. His configurable screensaver, *Organic Art*, marketed by Warners, has become a best-selling CD ROM.

To briefly describe the process by which Latham makes his work, he constructs virtual objects from various pre-defined shapes, usually horns and tusks. These are then modified by a program called Mutator which simulates evolutionary processes. Sets of numbers take the role of genes, while higher-level programs determine structures into which the resulting forms are fitted. The computer presents Latham with a set of nine forms generated from a starting image; he picks the one (or more, if there is to be virtual 'breeding') he likes best on purely aesthetic grounds, and this is used as the basis for the next generation.[6] The final pictures are merely the 'fruits' of what Latham calls an 'evolutionary tree' (Arnolfini Gallery, 1988–9: 13). It is the survival of the prettiest.

Latham calls these images 'sculpture', but at a conceptual level they are peculiar works of art: sculpture usually involves three-dimensional material, shaped by an artist and standing apart from the viewer. These computer visualizations do not necessarily have any such qualities. Higher-dimensional forms may be indicated, while fractals provide a visual expression of fractional dimensions (Todd and Latham, 1992: 51). Such images may be thought of as the shadows of higher-dimensional forms, their insubstantiality and complexity being the result of their materialization in a form that humans can appreciate. Their apparent scale and material are arbitrary. Viewpoint, lighting, colour and texture, while they may also be controlled by genetic formulas, are independent of the structures depicted. As digital forms, nothing about them is fixed and they provide the possibility for the viewer to interact with or even virtually become the 'sculpture'.

The manner in which Latham chooses to display these virtual artefacts, incidentally, is symptomatic of some of the contradictions in the ever-emergent field of computer art. Although plainly there is no original for these works, and the idea of reproduction has no meaning, Latham uses these forms to produce conventional art objects, making high-resolution colour photographs, printing them in limited editions and selling them through art galleries. So despite the radical potential of a medium which could place a Latham 'original' on every computer linked to the Net, rare and valuable objects are produced.

Latham holds to a simple faith that computer technology can produce beauty but has misgivings about the computer-aided genetic tinkering on which he wants his work to critically comment. It reflects, he claims, 'the computer age, man's genetic manipulation of nature and comments on the wanton destruction of the natural world' (Todd and Latham, 1992: 208). We shall see that this is a little disingenuous. But the pictures do have a creepy side: one of IBM's programmers, John Woodmark, described the works as 'stark but perfect, hovering in emptiness and lit only by a remarkable illumination that lights every crevice with an equal glow', and commented that he is glad that the creatures are confined behind the screen (Arnolfini Gallery, 1988–9: 26). The work is unsettling, partly because it impresses the eye with a convincing physical model which, however, departs in small details from what we normally see; the ray-traced rendering of virtual objects does not model how light dims with distance, nor how objects cast light on each other, nor penumbral shadows. Latham's images are uniformly sharp and often exhibit an awkward dislocation of elements, patterns and surfaces. But beyond this, Latham is dealing with a form of emergent order, the process by which very complex forms and systems can be produced by simple determinants. We see in the apparently intentional construction of these creatures, the signs of a non-human intelligence at work, and this may more than anything explain our unease.

To return to and further the argument about the transformation of modernism, let us briefly consider another moment of idealism, 1968. Aside from other events for which it is better known, 1968 was a crucial year for the development of computer art. *Leonardo* commenced publication; Jack Burnham's book *Beyond Modern Sculpture* was published, promising the evolution of sculpture into the production of autonomous, intelligent Galateas; Robert Mallary started to make computer-aided sculpture; and computer art was consecrated as a recognized force in contemporary art with major exhibitions in Berlin and at the Institute of Contemporary Art in London, while in New York, the Museum of Modern Art, caught up in the buzz about electronics, announced the end of the machine age.[7]

Yet at the same time, famously, this was both the high point of revolt and the beginning of disillusion – for Jack Burnham, writing a few years later, computer and technological art became part of a rebellion against the very idea of avant-garde progress (which he and others saw as an elaborate but decodable language game); a rebellion also against elite high art as a whole, and with it, many of the institutions and practices of the state and the economy (Burnham, 1971). Thinking of the permanently revolutionary and even iconoclastic art of modernism, and by contrast of the official bunkers in which it was housed, preserved and catalogued, he asked: 'is the ethos behind an invincible technology and a revolutionary art a reciprocal myth?' (Burnham, 1971: 41). Did both serve progress, did both unmake the world only to be unmade themselves as their products were assimilated by the system? There are various pernicious characteristics shared by science, technology and avant-garde art, argued Burnham: insane cycles of production and consumption, precociousness, fetishism, economic exploitation ... the list goes on (Burnham, 1971: 43). As high art drives itself into ever tighter circles of rebellion against itself, and, as a consequence, meaning is evacuated from it, the only way forward is via a great broadening of the practice of art, itself widely defined to include activist, radical acts – to make art democratic, local, natural. Technological art was linked, for a time, to a radical political and aesthetic project, promising the dissolution of high art along with the criminal institutions which sustained it.

Why is this moment of idealism important? Once again, there is the broken promise of liberal ideology. It is a salutary reminder that many of the social and political promises of the Net were already being made 25 years ago or more, but from a very different position than that taken today by the editors of *Wired*, with their unqualified faith in the beneficent effects of unregulated capitalism. It allows us to see clearly the emaciated nature of the radicalism promised today. And it is only one of a sequence of promises made about one gadget or another – even about television in its infancy – that it would be the conclusive social panacea, that it would finally bring true democracy. No technology can do that alone; there are, after all, powerful, structural interests which exercise their influence to assure that it cannot. These interests also tend to control the development and deployment of new technology. Yet, forever forgetting this, computer art and computer culture, eternal infants, cross the river Lethe once again, emerging wide-eyed, promising the Earth.

Modernism, then, was initially met with a critique which stressed the importance of human autonomy and initiative, emphasizing grassroots movements and emergent, egalitarian and non-prescriptive activity. When this ideal was defeated, both politically and culturally, some took refuge in virtual worlds where messy contingency could not check utopian construction. Only the form of the ideal was retained in computer culture, and in the anarchic Net promise of emergent culture – but the tendency was no longer to stress the potential of individuals but the inhuman force of the market. To think of this process in dialectical terms, a utopian, humanist modernism, defeated by political disillusion and economic recession, was replaced by a dystopian, anti-humanist postmodernism. Within the arena of a resurgent computer art, however, and this is only the forerunner of a more general trend, a dangerous synthesis has emerged – a utopian anti-humanism. And it is this complex, contradictory form which ensures the shuttling of the culture between utopia and dystopia.

Latham's work is part of a much larger trend which is engaged in examining and often proselytizing concepts of the 'post-human'. If there is something postmodern about computer culture, it is the attachment to the view that humanity has been somehow superseded, or at least that its time is beginning to draw to a close. These views may have radical or conservative aspects, whether it be Donna Haraway's politically correct cyborgs, or remade third-sex creatures beyond gender and biological hierarchy, or drones servicing the network, married to the great corporate collective Netmind (Haraway, 1991). While it may be a pleasant pastime to spin utopian or dystopian fantasies around the potential of these technologies, it is important to think not just about what is possible but what is likely. To do so, we have to examine computer culture today, and the forces which drive its transformation.

Is contemporary computer culture a haven of politically radical activity, anti-sexism, anti-racism and solidarity with Third World peoples? You can, of course, find all these worthy qualities within it – alongside much overtly fascist material – but looking at it broadly, the answer has to be no. Think only of that media preoccupation of yesterday, dildonics, or the high proportion of Net-searches devoted to finding pornography, or look at the advertisements in any of the gaming magazines. This is a culture which is both inconsequential and deeply conservative. Doom, the most popular PC game ever, is no more than an effective and violent shoot-'em-up – as is its successor Quake, the most hyped PC game ever. In both, aside from the obvious competence of the programming and the creation of a believable atmosphere, why such great success? Players are allowed considerable leeway to pursue their own tactics, but the basic idea is extremely simple; the enemies are demonic forces, single-mindedly devoted to the player's destruction. The player must become similarly single-minded and efficient in their destruction. Doom, Quake and now Quake II are very popular as networked games, creating between human players a particularly simple, all-or-nothing form of communication with virtual chainsaw, shotgun and rocket launcher.[8]

This is not to say that computer culture is worse than much other mass culture, or not at least much worse. As a form of mass culture it is, of course, largely commercially driven. Computer games create environments which are often hellish blueprints for the future or visions of the past. The virtual environments in which people currently dwell while working on computers are both more banal than such visions of heaven or hell, but also far from them in terms of meaning and action. (This was acutely visible in a recent performance by Stellarc, where the naked performer stood before the audience – a switching box strategically located over his groin – twitching and jerking in response to, or in control of, computer data and images thrown up on screens about him. But this fantasy or warning about a cyborg future was at odds with the Mac interface which so many people face every day. A little dog kept running – Stellarc's modified hourglass – while the computer threw up system errors. Everyone held their breath, waiting for the thing to crash.)

So it is easy to come up with characteristics of computer culture which are manifestly modernist, or which exhibit qualities that have traditionally been linked with modernism – at least, in postmodern critiques of the old tendency: sexism, racism, technophilism, single-minded and reductive narratives, and so on. It is much harder to refute the claim that the culture as a whole is somehow postmodern. But this difficulty, we should note, is to do with the nature of postmodern readings themselves; courting contradiction and relativism, there are naturally few explicit criteria against which to judge the suitability of a reading (there are implicit ones, of course, but we all know about those). So it is easy for postmodern critics to seize on some aspect of Net or computer culture and claim it for themselves.

And, indeed, we have seen that computer culture is a hybrid – modernist in its technophilism, its fetishization of functionality, and its steadfast ideology of progress; postmodern in its post-humanism, happy as it is to proclaim people the slaves of the new machines, as modernism (in practice, if not in words) had made people slaves of the old. In this new compact, postmodernism loses its quietism, the sense of resignation which made it content to conduct its wordy battles in the academy, letting the 'real' world (which, after all, had only a dubious status) take care of itself. But one of most the disturbing things about (what should it be called? – I hesitate to employ the term, after Kroker's use of the 'virtual class')[9] 'virtual modernism', is that the tasks Le Corbusier left to the Cartesian intelligence are now abandoned to the computer. Modernism's utopian content was often, and certainly in the schemes of Le Corbusier, much tied to order and hierarchy – and to authoritarian political schemes (McLeod, 1985). Yet within the haven created by such order, people were supposed to attain their potential and express their individuality: they would be equal but also diverse. People were the measure of the system.

The modified modernism of network culture is apparently far less attached to order, though that is implicit at least in interface design, but is also less tied to human values.[10] So the question becomes: how does the culture treat people, how does it make them act, what does it tend to make them become? Weiner's well-known work on feedback systems may provide a useful way of thinking of the human–computer 'interface' (Weiner, 1989). Net browsing in particular is just that – a feedback system in which human input is reduced to discrete mouse clicks, measured out in time and money.

What kind of human nodes does a collective, emergent Net culture require? Only dumb ones – or may it use smart ones, if people's cleverer actions cancel each other out, as lying is assumed to in opinion polls? This was an issue even in the early days of computer culture, when Robert Mallary wrote: '[Use] particular and bounded energy systems, such as human beings, as well as dispersed and ambient systems, such as crowds and technological and ecological forces. Finally, plug in (as it were) the total electronic environment as a source of the transductive [i.e. interactive] signals to shape, catalyse and energise the computer sculpture' (Mallary, 1969: 33). Here people act as mere elements, serving a wider cultural whole. Does such an order permit the self-consciousness of its agents to be expressed? Are we talking about the development of a democratic forum or a termite colony? Given all the hype about the radical political implications of the Net, what does it mean to extend the democratic project, to foster new communities or create a new public sphere when what it means to be human is rapidly changing, or, for that matter, when the mere presentation of individuals in cyberspace is arbitrary and constantly transformed?

To go back to Latham for a moment, we have seen that he produced an interactive screensaver employing the processes of Organic Art. The resulting forms are unforeseeable (by humans, obviously) and unrepeatable, and have a certain richness to them, despite certain cheesy pre-set elements. But the richness is all the computer's. Human interaction is limited to the picking of options, and that activity is highly controlled, limited and measurable. The point, then, is not whether emergent order requires dumb elements, but that humans surely can act as just that, becoming one of the inputs used in the construction, say, of complex marketing maps, charting the multidimensional relationships between different determinants, such as class, sex, race and the propensity to purchase lottery tickets.

Now if, as Burnham argued, avant-garde art and technology share certain undesirable features, perhaps they also carry within them solutions to the problems they create. Some people find Latham's images disturbing, and it may be that this response is not entirely irrational. The techniques of genetic computing can be turned to a variety of commercial applications – IBM's support for the artist was far from being pure altruism.

Genetic algorithms can be turned to generating spreadsheets as well as the images of strange creatures. They can be used, indeed, to generate and navigate through a vast number of financial plans. The advantage of this evolutionary approach over programs which merely optimize plans for a given scenario is that planners may recognize in the mutated results prospects for some outcome which they might never otherwise have considered. The immense advantage of running these plans virtually rather than using real resources and real labour is obvious. The computer, just as it is capable of mutating the elements of stick figures representing people, can juggle with prices, and investment, employment and layoffs.

Now this process simulates what economists term 'discovery', a matter that has always been an intractable problem for planned economies (Adaman and Devine, 1996). Given any particular scenario, through the competition of numerous private firms, many different alternatives are tried out, and the most successful is discovered and rewarded. In this never-ending process, the economy produces more knowledge about itself than is actually held by any of the individual participants, no matter how resourceful or well informed. Successful competitors may just be lucky, rather than frightfully bright and knowledgeable.

Obviously the genetic generation of plans can be used to simulate discovery, giving individual companies the ability to survey the field, letting them try out a multitude of virtual plans before committing themselves for real, so providing them with this market knowledge in advance. Of course, the program will only be as good as its economic model and the accuracy of the data it uses, but this is surely where networking comes in, making possible the automated seeking out and analysis of widely scattered data.

If discovery can be made virtually, then one of the main justifications for competitive markets is removed; a planned economy can exploit the maximization of resources which is produced by discovery, and, when it comes to judging the variety of possible plans, criteria other than profit maximization and shareholder benefit can be considered. The emergent order of the market can be emulated, and directed towards more human ends.

There is another consequence, however: if virtual discovery remains in private hands, firms will sooner or later have to build into their programs the calculations of competitors' virtual discovery programs; it could be that formerly rational courses will no longer be so, and that in this circumstance humans have little hope of hitting upon the courses that are made newly rational. What we may see, in other words, is a situation where raw computing power gives very great wealth to its owners, creating positions of fixed privilege which are even harder to shift than they are already.

To briefly sum up: the current manifestation of computer culture is not postmodern but an intensification and transformation of modernism, one which openly serves, not the needs of people, but those of 'higher' powers, in particular the universal power of the market. This has little to do with the technology as such – which might indeed hold out the prospect of radical change – but is determined by the current social hierarchy. The promise of the modernist ideal city has been transformed into the virtual hive, an ordered environment to which drones contribute their highly circumscribed signals to the emergent order of the market.

Why is technophobia so prevalent? Why do computers have to dress themselves up with cartoon characters or 'friendly' icons – why the triumph of the computing 'dummies'? It is surely not just computers' inherently non-human nature but that they have been and continue to be used for deeply anti-human ends: from the design and control of weapons to the calculations by which a bank decides how many workers it can afford to lay off. In the present circumstances, the ideology of capital dominates that of the Net and of computer culture in general. But it does not have to be that way: it is possible to see how, in a very different society, this non-human technology might be turned towards the most human of ends.

Notes

1 Among the large literature on this, see Buck-Morss, 1989.
2 For introductory reading on Le Corbusier, see Arts Council of Great Britain, 1987; Curtis, 1986; Moos, 1982.
3 Cocteau, 1926. On the 'call to order' see Green, 1987, and Silver, 1989.
4 The titles of such books as *Dos for Dummies* assure the reader that they are not alone in being foxed by computers and, indeed, that to be so is only normal, only human.

5 See for instance Tom DeWitt who argues that formalism, as in musical notation, may be extended to the visual arts, founded on algorithms and the procedural languages used in programming (DeWitt, 1989). See also Popper, 1993: 86–7. Similar arguments may be found in Franke, 1989. Another strand of activity, following the line pursued by D'Arcy Thompson and others since the 1930s, has been to study the structure of natural forms, and to write programs which generate similar shapes. It is visual rather than structural essence that is involved in such programming since there is no way of knowing this is not reverse engineering geared to reproducing the look alone.

6 See Todd and Latham, 1992: 64. Karl Sims has used similar programs to navigate a field of possible trees, which were again selected on aesthetic grounds. See Levy, 1992: 211f. There are other artists pursuing this kind of work, such as Yoichiro Kawagushi, who populates simulated undersea worlds with growing and mutating hybrid animal-vegetable 'creatures' with reflective skin. See Popper, 1993: 87–8.

7 See Institute of Contemporary Arts, 1968; Museum of Modern Art, 1968; Technical University, 1968.

8 A piece in *Wired* recently recommended Doom as an intuitive and reliable network environment in which people can compete and collaborate. See Goodwins, 1996: 37–8.

9 See Kroker and Weinstein 1994.

10 I suppose it is necessary to defend the use of such a term. See the discussion in Pinker, 1994: 413f.

References

Adaman, F. and Devine, P. (1996) 'On the economic theory of socialism', *New Left Review*, 221 (November–December): 54–80.
Arnolfini Gallery (1988–9) *The Conquest of Form: Computer Art by William Latham*. Bristol: Arnolfini.
Arts Council of Great Britain (1987) *Le Corbusier: Architect of the Century*, London: Arts Council.
Benedikt, M. (1991) 'Introduction', in M. Benedikt (ed.), *Cyberspace: First Steps*. Cambridge, MA: MIT Press.
Buck-Morss, S. (1989) *The Dialectics of Seeing: Walter Benjamin and the Arcades Project*. Cambridge, MA: MIT Press.
Burnham, J. (1971) 'Problems of criticism IX: art and technology', *Artforum*, 9(5): 40–5.
Cocteau, J. (1926) *Le Retour à l'ordre*. Paris: Stock.
Curtis, W.J.R. (1986) *Le Corbusier: Ideas and Forms*. Oxford: Phaidon.
DeWitt, T. (1989) 'Dataism', *Computer Art in Context: SIGGRAPH '89 Show Catalog* (Leonardo: suppl. issue): 57–61.
Franke, H.W. (1989) 'Mathematics as an artistic-generative principle', *Computer Art in Context: SIGGRAPH '89 Show Catalog* (Leonardo: suppl. issue): 25–6.
Franke, H.W. (1996) 'Editorial. The latest developments in media art', *Leonardo*, 29(4): 253–4.
Goodwins, R. (1996) 'Doom with a view', *Wired*, May: 37–8.
Green, C. (1987) *Cubism and its Enemies. Modern Movements and Reaction in French Art, 1916–1928*. New Haven CT: Yale University Press.
Haraway, D. (1991) 'A manifesto for cyborgs: science, technology and socialist feminism in the 1990s', in Socialist Review Collective (ed.) *Unfinished Business: Twenty Years of the Socialist Review*. London: Verso. pp. 257–76.

Institute of Contemporary Art (1968) *Cybernetic Serendipity: The Computer and the Arts*. London: ICA.
Kroker, A. and Weinstein, M.A. (1994) *Data Trash: The Theory of the Virtual Class*. Montreal: New World Perspectives.
Le Corbusier (1971) *The City of Tomorrow and Its Planning*, trans. Frederick Etchells. London: Architectural Press.
Levy, S. (1992) *Artificial Life: The Quest for a New Creation*. London: Penguin Books.
McLeod, M. (1985) 'Urbanism and utopia. Le Corbusier from regional syndicalism to Vichy'. Ann Arbor: UMI reprint of PhD, Princeton University.
Mallary, R. (1969) 'Computer sculpture: six levels of cybernetics', *Artforum*, 7(9): 33.
Moos, S. von (1982) *Le Corbusier: Elements of a Synthesis*. Cambridge, MA: MIT Press.
Museum of Modern Art (1968) *The Machine, as Seen at the End of the Mechanical Age*. New York: MOMA.
Pinker, S. (1994) *The Language Instinct. The New Science of Language and Mind*. London: Penguin Books.
Popper, F. (1993) *Art of the Electronic Age*. London: Thames & Hudson.
Poster, M. (1995) 'Postmodern virtualities', in M. Featherstone and R. Burrows (eds), *Cyberspace, Cyberbodies, Cyberpunk*. London: Sage.
Silver, K.E. (1989) *Esprit de Corps: The Art of the Parisian Avant-Garde and the First World War, 1914–1925*. London: Thames & Hudson.
Stoll, C. (1995) *Silicon Snake Oil: Second Thoughts on the Information Highway*. London: Macmillan.
Technical University (1968) *On the Path to Computer Art*. Berlin: Technical University.
Todd, S. and Latham, W. (1992) *Computers and Evolutionary Art*. London: Academic Press.
Weiner, N. (1989) *The Human Use of Human Beings: Cybernetics and Society*. London: Free Association Books.

PART 3

TERRITORIES

6 XS 4 All? 'Information Society' Policy and Practice in the European Union

John Downey

Undoubtedly, the dramatic development of information and communication technologies (ICTs) in recent decades has had a profound impact upon many areas of activity in the member states of the European Union. In the past few years, however, there has been a noticeable increase in policy emphasis on the importance of ICTs for the future prosperity and well-being of the Union which has largely grown out of a heightened awareness of the economic, social and political possibilities engendered by advances in networking technologies and a greater dissemination of ICTs throughout society at large. The development and implementation of ICTs in the European Union is now seen as a crucial goal of policy in a diverse area of activities. ICTs are generally viewed as a panacea by European Commission policy-makers and are thought to offer answers to seemingly intractable problems.

The importance of ICTs across the board has been recognized in policy debate by the increasing use of the concept of 'information society', replacing 'information technology' or 'information systems'. The concept of 'information society' has migrated from academic debate and informed journalism to policy arenas and one of the aims of this chapter is to track this migration and examine how academic debate has influenced policy-making.

The development of policy on the 'information society' has a number of different and highly complex aspects ranging from technical standardization and interconnectivity to data protection and security. While these subjects are of great importance they will not be covered here. Instead, this chapter will concentrate on examining policy thinking concerning the economic, social and political impacts of the 'information society' and will seek to question the rather breathlessly favourable prognosis which appears to be circulating more and more rapidly within the institutions of the EU. A fundamental contradiction exists between policies designed to improve the competitive situation of the EU *vis-à-vis* the USA, Japan and South East Asian economies and the goal of creating a more regionally and socially cohesive Europe. While it is difficult to see how this contradiction may be resolved, it is imperative that the negative potential of ICTs in terms of accentuating and deepening inequalities is more seriously assessed and addressed in the public sphere generally as well as, more specifically, by policy-makers.

The point of this is to develop a more realistic conception of the development of ICTs in the EU and to consider the potential impact of their implementation and wider acceptance in the future. To this end, a brief case study examining one of the leading 'superhighway' cities in the EU, Bologna, is offered since this will furnish evidence upon which more reasonable extrapolations can be based.

What is the 'Information Society'?

Webster (1995), though deeply critical of the recent proliferation of the concept of the information society in academic literature, policy-making and journalism, continues to use the concept, albeit with scare quotes. He does so because he does not want to dispute the increasingly important role of ICTs in the economic, social, and political spheres of advanced capitalist societies but equally does not want to associate himself with the new romantics of the 'information society'. Webster stresses continuity, not change. By this he does not want to deny that dramatic technological change has occurred in information and communication technologies since the end of the Second World War but rather seeks to emphasize the fact that the relations of production have not been fundamentally altered. Theorists such as Daniel Bell and Alvin Toffler associate the development of an information society with a radical break from the past. Bell sees this as the transition from industrial to post-industrial societies; Toffler sees this similarly but uses the concepts of 'second wave' to denote industrial societies and 'third wave' to denote post-industrial societies.

The naturalistic terminology is here important for it should warn us of a neo-technological determinism. Accompanying technological change is a move towards a better, more efficient, more just society. Webster quite rightly is intent on puncturing these assumptions by showing the inherently social character of technology and the complex relationship between technological and social change. Schiller (1998) goes a step further and sees the rhetoric of the rupture – the dramatic transition that the concept of the information society implies – as straightforwardly ideological in a traditional Marxian sense of a system of thought masking the nature of reality in the interests of the ruling class. Talk of rupture and the belief that the rapid introduction of information and communication technologies will result in the creation of an egalitarian society, according to Schiller, serves to obscure present power relations, thus providing them with sustenance. For Schiller, the US state and military-industrial complex act in harmony precisely to ensure that such a situation will not be brought about and increasingly utilize information and communication technologies for purposes of surveillance and propaganda (with the implication of conspiracy here Schiller is over-egging the pudding – the sincerity of the techno-boosters is not in doubt).

Although it is becoming unusual to see Marx's pejorative concept of ideology invoked so unashamedly in an academic context that has seen many shy away from making such grand epistemological and ethical claims, it must be said that Schiller does have a point. It is tempting to see the breathless theories of the information society as the expression of a particular class very much on the make. The youngish IT-literate middle classes possessing both the requisite economic and cultural capital to make good use of information and communication technologies have felt the wind of change sweeping through their open-plan offices (indeed, it is interesting to see how the architecture of the 'paperless' office is responding to the ideology of techno-boosterism). The accompanying utopian projections of the impact of ICTs often invoke Enlightenment thinkers' stress on the rebelliousness, nonconformity and individualism of that epoch. The concept of ideology becomes useful here analytically not because of the utopian hopes of the creation of a truly just society but because the practical suggestions of how the information society might be implemented actually serve to preclude the possibility of the creation of such a society. While Schiller sees Toffler as an important initiator of such discourses, one of the current high priests must be MIT professor Nicholas Negroponte.

In his March 1996 electronic epistle to *Wired* magazine Negroponte, director of MIT's Media Lab, takes issue with Jacques Chirac and his belief that the information superhighway represents a grave threat to the economic and cultural prosperity of *la grande nation*. The information superhighway is a threat, Chirac claims, because currently over 90 per cent of the transactions of the Internet take place in American English and thus the promise of the growth of networking threatens to Americanize the French. Nicholas Negroponte claims that the reverse is true. Far from exacerbating the marginalization of French-speaking cultures, the 'information superhighway' (an inappropriately modernist metaphor to describe a network but one that accurately reflects the teleological thinking of techno-boosters) will help them flourish. The 'global' network provides a space for the preservation of local and national identities. Negroponte cites as an example the work of Nichi Grauso in Sardinia who provides a browser available in 20 languages. Apparently this is an example of the intrinsically decentralized character of the Internet and leads to the conclusion that 'colonialism is the fruit of centralist thinking. It does not exist in a decentralised world' (*Wired*, March 1996: 42).

The problem with Negroponte is that he concentrates on the logic of the technology itself and excludes examination of the economic, social and political context in which the technology is used. The debate concerning the Americanization of Europe has a long history and contains many deeply problematical assumptions concerning the transfer of cultures via media of one sort or another. Media studies scholars will have a sense of *déjà vu* concerning Chirac's lament since similar arguments have circulated around Europe since the end of the Second World War – first about film, then television, and now about the Internet. Chirac's argument follows the logic of the market rather than the logic of the technology. Chirac does ignore, however, the greater potential for decentralization in networking than in either film or television and this is an important omission. The challenge is to examine the political economy of ICTs without ignoring the potential of using ICTs to mobilize resistance.

The liberal technologists of *Wired*, spearheaded by Kevin Kelly and Negroponte, draw on the rhetoric of the American revolution. Networking gives us the potential to begin the world anew. It is hard not to see such sentiments as the product of a particular, technologically sophisticated class desperately searching for societal solutions to apparently intractable problems. The vestiges of hope that had previously found expression in political movements are now projected on to technology:

> The methodical movement of recorded music as pieces of plastic, like the slow human handling of most information in the form of books, magazines, newspapers, and videocassettes, is about to become the instantaneous and inexpensive transfer of electronic data that move at the speed of light. In this form, the information can become universally accessible. Thomas Jefferson advanced the concept of libraries and the right to check out a book free of charge. But this great forefather never considered the likelihood that 20 million people might access a digital library electronically and withdraw its contents at no cost. (*Wired*, March 1996: 42)

Now this is a wonderful vision. The problem with Negroponte is that he sees advanced capitalist societies as already moving in this direction as though inequalities in these societies were getting smaller rather than bigger. For his vision to become true, everyone would not only need a computer and modem but would also have access to a free, reliable telecommunication network capable of handling large amounts of data. Negroponte needs to show how these and other inequalities are to be overcome. What does he suggest?

Negroponte's answer consists of liberalization and deregulation of telecommunications. Telephone companies need to be able to operate without the draconian regulations of the state. He mobilizes a deeply racist metaphor to argue for deregulation: 'the telecommunication business is regulated to such a degree that NYNEX must put telephone booths in the darkest corners of Brooklyn (where they last all of 48 hours) while its unregulated competitors will only put their telephone booths at 5th and Park Avenues and in airline club lounges' (*Wired*, March 1996: 42).

It appears then that universal access is only universal for some. Paradoxically, the ideology of information abundance is often accompanied by the claim that this reduces the need for state intervention. Perhaps the invocation of a Jeffersonian democracy is far more sinister than Negroponte consciously intended. It seems that the policies that Negroponte advocates to achieve his utopia will only ensure that the present inequalities in society deepen. This deeply contradictory argument seems to be suggesting that the only way to ensure universal service provision in the future is to deregulate in the present and let market forces have free play.

Now, Negroponte is an easy target. The reasons why an MIT professor would be a techno-booster are readily identifiable and understandable. What is, however, more disturbing is that this sort of thinking permeates public policy at national and international levels. The Internet is seen as the cure-all with telecommunication liberalization being the royal road to this information utopia. Not only is this an exercise in wishful thinking brushing away deep-seated societal problems, but the suggested means – liberalization – will serve to exacerbate these problems.

If in the first section of this chapter I have attempted to suggest that an 'information society' ideology exists and has gained a good deal of credibility in the public sphere, I am now seeking to focus on the impact of this way of thinking on the development of EC policy towards the goal of creating an information society.

Europe's Way in the Information Society

The 1993 CEC White Paper *Growth, Employment, and Competitiveness* identifies the growth of the information society as a key factor in the future economic and social well-being of the European Union. The White Paper adopts unhesitatingly the rhetoric of the rupture: 'a new "information society" is emerging' which will herald changes as dramatic as those which accompanied the first industrial revolution. The purpose of this section of the White Paper is to identify the measures which will speed up this process of transformation and yield maximum benefits for the citizens of the Union.

The development of ICT industries and the dissemination of ICT use is seen as central to the creation of new forms of employment. Not only will new industries and services be developed but the use of the ICTs will improve productivity and increase GDP in other sectors of the economy. Thus, while a certain measure of Schumpeterian creative destruction is to be expected with traditional occupations in both manufacturing and service sectors disappearing (previously largely immune from 'technological unemployment'), the overall impact on employment will be positive. On balance, it is suggested, more new jobs with new skills will be found than old jobs with old skills lost.

The White Paper does point to the potential difficulties of such a transition in society: 'the risk of exclusion, eg as a result of inadequate skills or qualifications and, more generally, the emergence of a two-tier society should not be underestimated'. While this is a laudable sentiment, it could be objected that the 'two-tier' society has already emerged and those seeking to identify those 'tiers' can draw on a wealth of empirical evidence in comparison to optimistic advocates of the information society. It also is unclear whether the policies suggested to overcome the risks of exclusion do not, indeed, underestimate the threats of economic, social and political exclusion.

The White Paper constantly looks over its shoulder at developments in the US and Japan. It is seen as very important for European industries to make up lost time on both their American and Japanese competitors in order to ensure that they are competitive in a global context. The key to ensuring this competitiveness is to liberalize telecommunications in the EU and thus foster competition, both between national operators who previously enjoyed monopoly positions in their domestic economies and, of course, with US and Japanese competitors. The provision of more services at lower prices is seen as a key factor in improving the competitiveness of the European economy, generating employment, increasing productivity, and GDP. But despite all of the beautiful words, what really seems to be driving policy is the goal of reducing leased-line costs to private industry, thereby enhancing their global competitiveness.

The role of the state in all of this is as a facilitator and regulator. The private sector is seen as the generator of change. A radical re-engineering of the fundamentally different relationship between public and private sectors is foreseen where the state instead of providing services simply regulates private provision. Citizens are seen as consumers of public services and delivery of services is dependent on payment and not on need. This is seen as part of the process of privatization of public services. Not only does this view the state as providing services according to use on a business–client model rather than providing services according to the needs of the citizens, but it also sees the proper home of these services as the private sector. What is noticeable here is how 'information society' policy neatly dovetails with the general privatization of all sorts of previously public provision in EU member states.

The underlying goal of policy initiatives is to create a 'common information area' in the European Union overcoming the economic, social and cultural differences of member states in order to improve economic competitiveness, and as such the information society initiatives need to be seen as part of a package leading to a single market. There are five necessary elements of a policy seeking to create such an area:

- diffusion of best practice of ICTs;
- creation of a legal, regulatory and political environment that will encourage private initiative while taking due account of the interests of citizens;
- development of basic trans-European telecommunications services;
- provision of training and development of human resources;
- encouragement of technology take-up by businesses and citizens.

The second element brings out the difficulties of the very idea of a common information area. The private sector is to be given a free hand to reduce telecommunications costs and make profits, yet this begs the question of the future of universal service provision. Some services are more universal than others; the common information area is only common for some. In order to advance this idea of a common information area the White Paper proposes to set up a 'very high level' task force with the objective of establishing the priorities in the development of the information society.

In December 1993 the European Council entrusted Commissioner Martin Bangemann, a former German liberal finance minister, to set up a group to report on the implications of the information society. The group consisted of, amongst others, representatives of the largest telecommunications and electronics firms in Europe. The report, perhaps unsurprisingly given the committee composition, concluded that the market should be relied upon to propel the EU into the information age. Measures should be taken to foster an entrepreneurial mentality and to provide for a common competitive regulatory framework. The report expressly stated that there should be no more public money, subsidies or protectionism. The market, in other words, provides the solution while the state's job is simply to encourage the market through deregulation. If we follow these suggestions the Bangemann Report claims 'we will all win in the long run'. While the Bangemann Report recognizes that the chief risk of the information age lies in the widening of the gap between the information haves and have-nots it supplies no convincing recommendations for how this is to be avoided. As a concern it certainly plays second fiddle to the drive for productivity gains through networking. What is really important is catching up with the US and the emerging economies of Asia in terms of productivity. It is difficult to see how an essentially free market policy will result in all of us being winners.

The Bangemann Report, despite its adoption of the hype of the techno-boosters, recognizes that there has been a slow take-up of demand and supply in the EU and suggests that projects demonstrating applications are funded in order to act as a catalyst for change. One of these applications is the creation of city information highways. This is interesting as it implicitly admits that rural areas will struggle compared to cities because of their unattractiveness to telecommunications operators. It is also the only application mentioned that has possibilities for creating citizen involvement rather than simply making businesses more competitive or delivering services more efficiently. In the final section of this chapter I will examine the progress that has been made in creating city information highways by looking at one of the most successful cities in the EU, Bologna.

While both the 1993 White Paper and the 1994 Bangemann Report can be safely accused of a naive technologically driven liberal utopianism, the forums, groups and studies that have followed in its wake have provided a more sober assessment of the impact of the information society.

Set up in 1995 the High Level Group of Experts (HLEG), chaired by Luc Soete and including Manuel Castells, a leading academic authority on the economic and social implications of the information society, delivered an interim report based upon first reflections in January 1996. The title of that report, *Building the European Information Society for Us All* (HLEG, 1996), is significant as it indicates the first signs of resistance to the 'no-losers' scenario set down by the Bangemann Report. While not contesting the potential of ICTs to act as the motor of economic and employment growth through boosting productivity, the interim report examines the likely impact of ICTs on social cohesion both within and between European regions. Until this report the social and political implications of the information society had been glossed over by an insistence on the curative properties of the free market. The report is also significant in its implied critique of the Bangemann Report's neo-technological determinism: 'the technology in itself is neither good nor bad. It is the use which human beings make of any technology which determines both the nature and extent of the benefits. Moreover, these do not accrue automatically to everyone in society' (HLEG, 1996: 10).

The HLEG raises the question of the employment impact of ICTs. In the past decade or so employment growth in the EU has largely been confined to service industries and until recently service industries were seen as labour intensive and insulated from technological unemployment. With the increasing implementation, however, of ICTs the possibility of technological unemployment in service industries, they argue, will become a reality. A process of 'creative destruction' will result in a rationalization of services, particularly at the front end, with customers and citizens becoming able to get what they want electronically and without direct face-to-face human interaction. Whether job losses ensue is dependent upon the private and public sectors' attitudes. Do they sack or do they redeploy their labour forces in order to provide a higher level of service? The HLEG recommend the latter approach, but in a climate where the ideology of the 'lean state' is gaining widespread acceptance it is hard to believe that this is the course of action that the public sector will follow.

The deployment of ICTs, argues the report, will strengthen the trends towards 'globalization'. The 'annihilation' of distance paradoxically makes the relative advantages of particular locations more important. This is especially true for lower-skill employment and there is already ample evidence of the growth of outsourcing from regions where labour rates are higher. Because of problems of translation it is likely that 'clerical outsourcing' will be mostly a national phenomenon but for other areas of activity the ICTs' benefits for command and control functions will give added impetus to international outsourcing.

Similarly, ICT implementation would seem to compound recent post-Fordist tendencies in advanced economies. Large firms are 'downsizing' and subcontracting work that was previously carried out in-house. Improved communications between enterprises would make an acceleration of this tendency more likely. Likewise the casualization of the labour force, the rise of short-term contracts, zero inventories and piecework will all gain added impetus from more widespread use of ICTs. Traditionally, trade unions have derived their power from their ability to organize workers at the point of production. When the points of production multiply in the case of post-Fordism and in the rise of telework, unions will have to organize and agitate electronically. Judging by the difficulties unions have encountered in the past in organizing an increasingly casualized labour force, the prospects for virtual class resistance do not look promising.

The decreasing demand for people with lower skill levels is being reinforced by greater use of ICTs. The need for a literate and numerate workforce is being joined by the need for computer literacy. Workers without access to retraining possibilities will be more and more confronted by economies which have little need for them. Younger, better-educated and more computer literate people will thrive in the information society while others will face an even harder struggle to hang on to their jobs or find new ones. This raises the question of how the improved economic growth likely to be engendered by ICTs will actually be distributed in the information society. Since 1979 economic inequality has grown in the UK and it appears unlikely that, assuming a genuine reforming social democracy does not wrest ideological leadership from the 'New Right', this gap will close. Indeed, the opposite is more likely.

Post-Bangemann there has been a subtle shift in emphasis in EC policy-thinking, concentrating on making the information society as inclusive as possible. This is largely down to the work of the HLEG. It is worth examining how this shift of policy emphasis is related to academic work in the field. For my purposes, it is appropriate to look at the work of Manuel Castells, one of the EC's experts.

The Informational Mode of Production and Inequality

Although Castells accepts the vocabulary of the rupture, of a dramatic, revolutionary change sweeping through advanced capitalism as a result of developments in ICTs, he comes to significantly different conclusions from the techno-boosters who promise that there will be no losers.

The effects of the developments in ICTs are pervasive, Castells argues, i.e. they are apparent throughout advanced capitalist societies. ICTs are 'transforming production and consumption, management and work, life and death, culture and warfare, communication and education, space and time' (1994: 20). Thus, he claims, one can talk of an informational mode of production which is felt in manufacturing and service industries alike and is the driving force of societal change in a new phase of capitalism.

These changes are accompanied and augmented by globalization – the shift from a world economy, such as that described by Wallerstein, to a global economy which 'works as a unit in real time on a planetary scale'. It is important to recognize here that Castells is not asserting that all zones and regions participate equally in this global economy. Quite the reverse, in fact. He argues that it allows metropolitan centres even greater economic and political importance as it enhances their ability to command and control and further isolates much of Africa as capital bypasses it in search of return. Whole regions find themselves out of the loop.

Castells analyses the process of 'suburbanization' in advanced economies which is part of the dialectic between centralization and decentralization created by the shift in advanced economies towards an informational mode of production: 'the production of surplus derives mainly from the generation of knowledge and from the processing of necessary information (Castells, 1989: 146)'. Despite the claimed developmental logic of information/service industries (decentralization) using ICTs, Castells argues that communication technologies have had no direct effect on the location of industries, with metropolitan areas still dominating. What is crucial is the logic of the market not of the technology itself. There has, however, been a decentralization of service industries within metropolitan areas (the creation of so-called 'edge cities'). Companies take advantage of cheaper rents, building possibilities, better working environments, etc.

The creation of a global economy, Castells contends, is at the expense of national economies. The interests of capital are diverging from those of nation states. This transition is best characterized by three interdependent broad processes:

- reinforcement of the metropolitan hierarchy;
- decline of traditionally dominant manufacturing industrial regions in advanced capitalist economies as an increasingly mobile capital seeks out comparative locational advantages, particularly lower wage rates and taxes;
- emergence of new industrial regions benefiting more or less directly from the decline of traditional manufacturing centres.

The likely consequences of these processes are greater inequality between regions and cities, and greater inequality within regions and cities.

The greatest economic benefits are likely to accrue to 'first-mover' regions and cities, those cities where command and control functions are exercised and where the logic of ICT deployment is obvious. Developing regions will benefit economically from capital's greater mobility (although they will be dependent upon the favour of metropolitan centres) while traditional industrial regions will struggle to adjust to the changed terms of trade of the global economy.

The 1996 EC Green Paper *Living and Working in the Information Society: People First* places emphasis on the goal of achieving an economically and socially cohesive Union. It sums up what has been achieved to date: 'progress towards convergence in income per head between Member States has been positive but slow, but disparities between regions within the same

Member States have tended to widen over time' (Commission of the European Communities, 1996: 43). This would seem to support Castells's claim that we should consider regional and city economies in the context of globalization rather than national economies since some regions will profit and others will lose out as the process of globalization takes hold. Disparities between regions appear to be growing and are likely to do so in the future with the increased importance of ICTs for businesses. In a communication to the European Parliament *Cohesion and the Information Society* (CEC, 1997) the Commission notes wide disparities in telecommunications provision between core and cohesion regions in the EU. (The cohesion regions are Ireland, Greece, Spain, Portugal, and some regions in southern Italy.) A number of different indicators were used and what follows is simply a selection which appears to confirm the argument concerning polarization.

In the cohesion regions the number of Public Switched Telephone Network (PSTN) faults per 100 lines is almost three times as great as in core regions (32.6 as opposed to 11.3). Integrated Services Digital Network (ISDN) covers only 32.5 per cent of the cohesion regions compared to 85 per cent for the core regions. Cable TV covers 54.4 per cent of core regions and 8 per cent of cohesion regions. In 1994 the cost of residential subscription for a telephone as a percentage of GDP was twice as expensive in cohesion regions as in core regions. In core regions there was in 1995 one Internet host per 200 people compared to one per 1,300 people in cohesion regions. In 1995 there were 15.2 PCs per 100 people in core regions compared to 3.1 per 100 in Greece, 8.6 in Spain, 6 in Portugal, and 8.9 in Italy.

What these figures indicate is not just that the cohesion regions lag significantly behind the core regions in terms of ICT development but also that the barriers to development are also very much higher in the cohesion regions. Bearing in mind that the benefits of ICT use are likely to accrue to first mover regions it seems likely that the disparities in ICT use will result in even greater income inequality between core and cohesion regions in the future.

The greater inequality between regions and cities is being matched by greater inequality within regions and cities. De-industrialization has seen a decline in demand for traditional blue collar work in many regions of advanced capitalism. While this has partially been offset by increasing demand for low-skill white collar service industry jobs these occupations tend to be characterized by low pay, poor conditions and job insecurity. Within these cities it is those people who already possess economic and cultural capital that have the ability to adjust to the demands of a changing occupational structure.

The UK may be used as an example to highlight the labour market changes occurring in many advanced capitalist societies. It is forecast that between 1993 and 2001 the labour force will increase by 1.2 million with a growth in employment of 1.5 million. Most of these jobs will be taken by women aged between 25 and 50 years, representing a further 'feminization' of the workforce. There will be an additional shift away from agriculture and manufacturing towards the service sector although this will be accompanied by a decrease in secretarial and clerical jobs (which had grown steadily for 30 years or so). This may be directly attributed to the impact of information and communication technologies. There will be a decline in full-time jobs and an increase in part-time and self-employment. In terms of changing skill requirements there will be a decline in demand for low-skilled manual occupations except for personal and protective services (here an increase of 523,000 is forecast). The largest employment growth will be in more highly skilled, white collar occupations, reflecting organizational and technological changes. These changes will favour higher-skilled, white collar, non-manual employees.

These figures amply illustrate Castells's thesis of an increasing polarization of the workforce. People with economic and cultural capital will prosper as they are able to adapt to technological and organizational changes. Employment growth elsewhere is confined to low-skill and low-status service occupations which will largely be taken up by women. The only category expected to see a growth in employment in a traditionally male area will be security services. These employment changes would seem to indicate that income inequality will increase, reflecting a growing division between professional and skilled workers and the rest.

Castells argues that this will lead to a reinforcement of dual cities, a concept he borrows from the Chicago School's work on urban societies. He contends that 'the informational economy has a structural tendency to generate a polarised occupational structure, according to the informational capabilities of different social groups. Informational productivity at the top may incite structural unemployment at the bottom or downgrading of the social conditions of manual labour, particularly if the control of labour unions is weakened in the process and if the institutions of the welfare state are undermined by the assault of conservative politics and libertarian ideology' (1994: 29–30). Perhaps it is here worth remembering the individualism and libertarian ideology of the techno-boosters (even when they see themselves as representatives of the tradition of social democracy) and their impact on EU policy development.

What Castells is implying here is that significant sections of the working class in advanced capitalist cities will be objects of the information society rather than subjects and agents. The weakening of labour movements is part and parcel of the creation of a global economy.

Mike Davis's work on the history of Los Angeles is helpful in contemplating the relationship between physical and electronic surveillance in dual cities. Rather than see clear-cut distinctions between real and virtual cities it is important to see them as intertwined. Both physical and virtual space are a reflection of the society which creates them. The post-liberal fortress architecture of Los Angeles has seen a gentrification of some of the zones of the inner city accompanied by 'a proliferation of new repressions in space and movement' (1992: 223) representing an 'unprecedented tendency to merge urban design, architecture and the police apparatus into a single comprehensive security effort' (1992: 224). Davis argues that security has become a 'positional good', with the wealthy buying private protection from the disenfranchised. These changes in physical architecture are matched by the electronic architecture of cities: 'the privatisation of the architectural public realm is shadowed by parallel restructurings of electronic space, as heavily policed, pay-access "information orders", elite databases and subscription cable services appropriate parts of the invisible agora' (1992: 226). The pseudo-public spaces within fortress cities are both physical and electronic and consequently the defence of the idea of public space and the public sphere needs to have two fronts. However, in comparison to the defence of physical space the defence of electronic space has received little attention.

Political resistance, Castells suggests, is more likely to come from urban subcultures, which are at once disaffected and yet can use their acquired cultural capital to resist. The problem with this quite realistic assessment is that these political groupings and movements (the green movement, feminism, animal rights, etc.) have often ignored class politics and so are not geared up to protect the interests of those who will be excluded from the information society.

Such subcultural resistance could be facilitated, Castells argues, by local government. He recognizes their comparative lack of power and suggests three transformations necessary if resistance is to be effective:

- citizen participation;
- cooperation between local governments on a European level;
- reaction to the social trends associated with globalization.

While this is essential, a more sanguine assessment of the possibilities of mounting effective resistance needs to consider:

- financial constraints placed upon local governments by the centre in a climate where the commitments of the state in many areas are being minimized;
- the costs of supplying on-line services as well as paper services to citizens. This involves a duplication of effort, with dual delivery;
- lack of experience in dealing with issues of telecommunications since previously this was largely a national issue and did not fall within the remit of local government;
- a democratic deficit, felt particularly keenly at local level, where local government is often not understood or is regarded as being powerless by citizens.

What is absent from Castells's work is a theorization of how these local initiatives will relate to the political process generally and the public sphere in particular. Here, the ideas of Jürgen Habermas may prove helpful in examining the possibilities of electronic political debate and mobilization.

Public Cybersphere?

Habermas's early work on the public sphere has received much attention in English-speaking media studies since its translation in 1989. Although Habermas has been criticized from a number of quarters, the concept of the public sphere has achieved a certain amount of acceptance by 'centre-left' intellectuals because it supplies a straightforward theoretical framework which can be used in present political struggles concerning the 'privatization' of the public sphere and the increasing concentration of media ownership. Thus far, however, because of the limited recognition of the implications of convergence, telecommunications and computing technology have been peripheral to this debate.

While agreeing that the development of a bourgeois public sphere in the eighteenth century was dependent upon exclusions of class, gender and race, Habermas maintains that the liberal ideal of rational debate engendered at this time was an important moment in the development of democratic values. Such debate, he argues, appears to have little place in twentieth-century liberal democracies as powerful private interests come to dominate the public sphere and politics becomes more akin to public relations than to rational dialogue.

Thirty years on from his original work on the public sphere Habermas seeks to revisit and revise the concept in the context of contemporary politics in liberal democratic states. His revisions offer the prospect of resistance and reinvigoration of the public sphere in contrast to his earlier deliberations, which were reminiscent of Adorno's thesis of the totally administered society. To be sure, Habermas argues, when one examines the control exerted upon the public sphere by the private interests of, for example, media moguls one must be sceptical about the ability of the bodies which make up civil society to influence political discourse.

Habermas argues, with a good deal of justification, that there has been a depoliticization of the public sphere:

> the personalization of questions of policy, the mixing of information and entertainment . . . the fragmentation of contexts lead to a syndrome that supports the depoliticization of public communication. (1992: 456; my translation)

However, this image of a closed public sphere is a result of seeing the public sphere as a static entity, as an 'Öffentlichkeit in Ruhezustand' (the public sphere at a standstill). Once the public sphere is seen as dynamic and as subject to crisis, one can plausibly argue that gaps open up which can be occupied, or at least influenced, by previously ignored actors. This is made possible by the ideology of the mogul-dominated mass media which presents itself as being responsive to public opinion. When crises occur and groups in civil society manage to mobilize popular resistance, Habermas argues that these groups can exploit the cracks that open up in the public sphere and play 'a surprisingly active and consequential role' (1992: 460). The prime example of this, Habermas contends, is the green movement in Germany.

When contemplating the possible role of a public cyberspace it might be helpful to see it as an adjunct to, rather than as a replacement of, the market and mogul-dominated sphere of mass communication and see its importance in terms of the influence that it might be able to exercise on the mass media in times of crisis. Certainly the economic and cultural barriers to participating in public cyberspace have meant that, so far at least, the participants in dialogue in virtual coffee houses bear many similarities to the white, bourgeois men of the real coffee houses in the eighteenth century. This is not to deny, however, that cyberspace offers a new place for discussion, dissemination and opposition for actors in civil society from which, to adopt Gramsci's terminology, assualts can be made in the war of position. Of course, these opportunities can be grasped by the Right as well as the Left, which means that the consequences for the public sphere as a whole are not necessarily progressive or emancipatory. Perhaps the surest conclusion is the modest one that we can expect a certain destabilization of the public sphere as a consequence of increased use of new information and communication technologies by non-governmental organizations and the like.

This conclusion contrasts with much of the hype which sees ICTs as a tool for democracy. The conversations that take place in cyberspace may be convivial; they may also be brutal and reactionary. The key is to conceive of cyberspace in the same way as other media of modernity have come to be regarded – as a site of contestation. Once we have moved on from thinking about the in-itself of technology, of whether the form itself is progressive or reactionary, we can start on the project of attempting to make cyberspace what the mass media largely are not, i.e. inclusive and participatory.

The major stumbling block to this is, of course, the belief expressed in international and national policy that the 'communications revolution' ought to be driven by the private sector and the accompanying measures of liberalization of telecommunications. This may serve business interests in the EU in their competition with the USA and Japan but it certainly does not serve the interests of citizens. Given this free market approach at EU and national levels, however, it appears that in the short term, at least, it is left to the state at local and regional levels to promote cyberspace as an inclusive, diverse and participatory arena. Since local government cannot greatly influence the development of private networks its room for manoeuvre is circumscribed though not unimportant. It largely concerns three areas of activity: the provision of public terminals; the provision of free and low-cost on-line services; and the provision of training. Paradoxically, much of the funding for these sorts of activities is provided by the EC, which points to the present contradiction in EU policy between serving the interests of business, on the one hand, and the interests of citizens, on the other.

City Superhighways in Europe: a Brief Case Study

In 1996 the City of Stockholm in an attempt to promote the information society and their place in it started the Bangemann Challenge. Ten awards were made to EU cities which had made most progress in the application areas identified by the Bangemann Report as most likely to kick-start the information society in Europe. This was followed up in 1997 by a Global Bangemann Challenge open to all cities with a population of over 400,000. Amsterdam received a special award for its pioneering efforts. Since activities in this city were discussed by Bergman and van Zoonen in Chapter 4 of this volume, here I will concentrate on Bologna, which received an award for its public information initiative.

Many factors help to explain why Bologna is one of the first movers. Bologna is a relatively rich European city, so it has a public telecommunications system which is far superior to those available in cohesion regions of the EU. It has a well-respected university with a large number of students and many people working in professional occupations and skilled jobs. These people are more likely to use ICTs than other groups. They act as a catalyst, providing a critical mass of users. Bologna has also traditionally had a communist-controlled city council which sees itself as playing an active role in governing economy and society rather than simply as a regulator of the free market.

The Iperbole civic network operates in the municipality of Bologna and the surrounding metropolitan area, Emilia-Romagna. The total target population is in the order of 900,000. It was started in January 1995 and financed by the city council with additional EC funds. The network provides citizens with free-of-charge e-mail, electronic access to the council and other public bodies, free connections and software. A full Internet service is made available free of charge to certain disadvantaged groups while others receive subsidized access. There are currently 35 newsgroups acting as an interface between council decision-makers and citizens, covering such issues as urban planning, social problems, culture and the environment. Latest figures available (July 1997) show that about 1.5 per cent of the total population are registered users (11,000 users). Additionally, 700 organizations, 63 schools, and 130 municipal offices provide information services. There are something like 14,000 information files. The information is accessed from 30 public access points as well as from dial-ups from work and home. There are 35 new users per day with a two-month waiting list for connection. The number of users exceeded 16,000 by the end of 1997.

These are undoubtedly extremely impressive figures in comparison to other digital cities in Europe. However, it is important not to be carried away, and further research on users is required. It is likely, for example, that the people using the network largely come from the professional middle classes and the student population and so the network is effectively subsidizing relatively advantaged citizens. This is not to criticize the idea of a civic network but rather to show the distance that needs to be travelled in order to make the network a genuine public space. Perhaps it is helpful to think of the development of digital cities as a two-stage process. The take-off relies on the enthusiasm and commitment of local governments and the educated activist section of the middle classes but the development of an electronic public sphere depends on reaching out to groups traditionally excluded, both economically and culturally, from information technology. This second stage is the most difficult and has yet to be realized.

While Bologna has made progress on a practical level it has also set the agenda on a political and theoretical level. The council sees itself as a primary actor in the development of ICTs rather than simply having a secondary role in the regulation of the free market. The provision of public access, free connectivity, user-friendly interfaces and training is seen as essential to enable citizens to exercise their 'communicative rights'. The local state is thereby promoting positive liberties (freedom to) as well as ensuring the traditionally liberal negative liberties (freedom from) of citizens. The application of such thinking to the electronic realm has much to do with the political traditions of the city council and places it apart from much other local government in the EU and the increasing trend in the EU to see the state as primarily a regulator while providing a minimal safety net for those who are excluded by the free market. The influence of Bologna, therefore, needs to be felt not only at a practical level in terms of disseminating best practice but also at a political and theoretical level. Indeed, it is likely that without the widespread adoption by local government of the principle of positive communicative rights of citizens digital cities will remain pseudo-public spaces.

Conclusion

While it is likely that the greater use of ICTs will have significant benefits in terms of productivity, GDP growth and employment, it is also probable that these benefits will not be equally distributed. Inequalities between core and peripheral regions will grow as core regions increase their grip on the global economy; inequalities within cities will widen as new highly skilled jobs are created while low-skilled manual jobs are destroyed. When these economic and occupational trends are coupled with the liberalization of telecommunications in the EU, we are likely to see a reinforcement of dual cities, a situation where significant numbers of people are disenfranchised politically and excluded economically. Given the EC's faith in a free-market-driven information society and its emphasis on improving the competitiveness of EU industries compared with the USA and Japan, it is difficult to see policies emerging in the short term that will resist this process. Regulation of telecommunications and the ideals of universal service very much play second fiddle to the goal of competitiveness. It is then left to nation states and to local government to attempt to guide this process for the benefit of citizens, although obviously their hands are tied by the overarching EC policy of liberalization. Not only is the sphere of action of local government limited but there is also a plethora of other constraints, both economic and cultural in character. Having said that, given the current political climate, it is likely that it is at the grassroots level (if at all) that resistance will emerge even if the chances of success are not good. On a practical level, free access to public terminals, training and e-mail together with user-friendly interfaces are all to the good. However, just as importantly, the dominant discourses of the 'information society' need to be challenged both in the public sphere and in the policy arena. It is here that Bologna's initiative is so impressive because practical steps have been accompanied by a demand for citizens to have positive 'communicative rights'.

While the Internet and World Wide Web as they stand may be used by activists to organize and disseminate, this is made possible by their ability to adapt to a pseudo-public place and is dependent upon the possession of both economic and cultural capital on their part. For a genuinely public space to emerge a radical change of policy at EC level is required together with a considerable change in both economic circumstance and ideology of states at national and local levels. The prospects for this are not encouraging.

References

Bangemann Report (1995) *Europe and the Global Information Society: Recommendations to the European Council.* Brussels: European Commission. (http://www.ispo.cec.be/infosoc/backg/bangemann.html)
Castells, M. (1989) *The Informational City: Information Technology, Economic Restructuring, and the Urban Regional Process.* Oxford: Blackwell.
Castells, M. (1994) 'European cities, the informational society, and the global economy', *New Left Review*, 204: 18–32.
Castells, M. (1996) *The Rise of the Network Society.* Malden, MA and Oxford: Basil Blackwell.
CEC (1993) *White Paper on Growth, Competitiveness and Employment: The Challenges and Ways Forward in the 21st Century* (Com 93 700 final).
CEC (1997) *Cohesion and the Information Society* (COM 97 7/3) (http://www.ispo.cec.be/infosoc/legreg/docs/cohes1.html).
Commission of the European Communities (1996) *Living and Working in the Information Society: People First* (Com 96 389 final). Luxembourg: Official Publications Office of CEC).
Davis, M. (1992) *City of Quartz: Excavating the Future in Los Angeles.* London: Vintage.
Habermas, J. (1989) *The Structural Transformation of the Public Sphere: an Inquiry into a Category of Bourgeois Society.* Cambridge: Polity Press.
Habermas, J. (1992) *Faktivität und Geltung.* Frankfurt-on-Main: Suhrkamp.
HLEG: High Level Group of Experts (1996) *Building the Information Society For Us All* (http://www.ispo.cec.be/hleg/Building.html).
Schiller, H. (1998) 'Striving for communication domination: a half century review', in D. Thussu (ed.), *Electronic Empires: Global Media and Local Resistance.* London: Edward Arnold.
Webster, F. (1995) *Theories of the Information Society.* London: Routledge.

7 Beyond Infrastructure: Europe, the USA, and Canada on the Information Highway

Leen d'Haenens

A government's policy with regard to the mass media can be expressed according to different degrees of state intervention: freedom of opinion and state control are the two poles between which all policy options are to be found. Traditionally, the only choice has been between three options: free speech, a partially regulated framework, and 'the voice of the government' as was seen in the early days of television. In Western democracies policies regarding the print media are based on the liberal 'freedom of expression' paradigm. Conversely, the partially regulatory paradigm still dominates policy as far as the audiovisual media are concerned.

Cyberspace looks set to become the most pervasive cultural mass medium of all time. In fact, it could be a revolution in its own right, on a par with the microcomputer revolution which spawned it. It can potentially reach audiences on a scale unprecedented by any other medium. Such large-scale intercultural communication would have far-reaching social and cultural implications, and not everybody stands to gain from it. This promises to provoke forceful opposition along the usual protectionist lines. Still, the information superhighway could become the carrier of choice for existing products and services: news, sporting events, films, drama, video games, videoconferencing, databases, etc. It is also a medium suitable for interactive programs and services based on direct participation of the end-user. Almost all non-interactive content that will be delivered to the consumer via the information highway already exists and is currently carried by other media. Above all, the information superhighway promises the convergence of delivery of traditionally distinct forms of media content.

First in the USA and (to a lesser extent) in Europe and Japan, vertical and horizontal alliances are being formed between sectors that used to be clearly separate (each with its own regulatory framework), such as broadcast media, telecommunications and a variety of computer-based business activities. These are the characteristics of an information society in which multimedia, scale amplification and mega-mergers of all kinds are the order of the day (Bouwman, 1994: 2). One direct, global consequence of this blurring of boundaries between formerly distinct media is highly problematic for policy-makers. Effective mechanisms embracing new information technologies as a whole seem extremely difficult to find. Because of the lack of integrated policy tools, policy-makers are failing to tackle the problem on a macro level, tending instead to look at it as a set of discrete aspects and trying to find meta-level tools that they hope will be effective on specific issues such as privacy and the protection of information, universal access, interoperability, standardization, etc. Apart from this central issue – blurred boundaries between formerly well-defined media and the corresponding policy frameworks – we will see that national, supranational or regional policies all vary according to the geopolitical, cultural-linguistic and, last but not least, business climates and realities in which they are developed and implemented. Trying to understand the differences and similarities in policy options in the regions under scrutiny, we find that some constant features (including mistakes) of past audiovisual media policy shed a great deal of light on those policy options currently being adopted by governments with regard to information and telecommunications.

Aims and Objectives

First of all, this chapter intends to compare some major differences and similarities in the approaches adopted by governments and industrial partners throughout Europe, the USA and, to a lesser extent, Japan, in order to implement the information society (see also d'Haenens, 1998) – with the information superhighway seen as one of its driving forces. A look at the relevant literature (not only learned journals and books, but also government reports and official documentation of supranational organizations such as the European Union) shows that the overall goals are the same, although the methods adopted by the various governments may vary with respect to (1) creating the best possible infrastructure and (2) utilizing the infrastructure for economic and social benefit. Key issues throughout these regions are societal and political aspects such as the impact on the job situation in the short, medium and long term, intellectual property rights, cross-ownership of media and antitrust regulation, privacy, censorship, security of electronic information, and universal access. Such issues are crucial. To do them justice we would need to go way beyond the limited scope of this chapter.

The chapter looks into the Canadian perspective on the information superhighway, since options taken in Canada can offer interesting perspectives on the potential path to be followed within the European Union. Apart from the necessary technical infrastructure (networks, terminals), one cannot ignore the fact that the information superhighway's vitality will be directly dependent on the content on offer. We hardly need to point out that in the past many promising technological achievements went the way of the dodo because they failed to fire the imagination of the buying public. One may therefore contend that the information superhighway's success will be greatly dependent on the cultural sector as a prime content provider and indeed, whether the highway delivers benefits for society as a whole. So, beyond financial and technological issues there is a need for basic principles governing content supply. The Canadian government has developed the following provision concerning content:

> Content should reflect the international diversity of perspectives and languages for the benefit of a majority of users. The emphasis in the supply must be on openness instead of concentration; on diversity rather than one single perspective.

What is on offer on the information superhighway should not be dictated by the G7 countries' narrow and often protectionist self-interest. According to the Canadian vision, what is interesting about the information superhighway has more to do with communication and cooperation than with information. This is the basic difference from the American stance, which emphasizes infrastructure and raw data sharing. Canadian government documents indicate that cyberspace should not be a mere hub regulating the flow of data criss-crossing the globe, but that it should be promoted as the meeting point for those various dynamic communities that make up Marshall McLuhan's global village – a space where creative minds cross-fertilize.

USA, Europe, and Japan: Differences and Similarities

Before going any further, we need to briefly assess current trends in the US, Japan and Europe, some of which are diverging while others are similar. In addition, we shall look into the changing role of the government and the difficult task of finding the right balance between freedom and regulation with regard to the establishment of standards for new information and telecommunications technologies. One of the key policy goals is to develop open standards: 'The goal is to come to unambiguous, uncontested standards' (Libicki, 1995: 73) in order to exploit the threefold economic advantages of compatible standards ('they may lower supplier costs, increase consumers' willingness-to-pay, or alter the competitive dynamics among suppliers' (Lehr, 1995: 125)). We will, therefore, also take a look at the standardization process in the USA and Europe as well as, to a lesser extent, in Japan.

Table 7.1 (Longhorn, 1995: 8) illustrates the differences in these three major economies, taking into account the following four criteria of information content, network structures, applications and software, and the 'people' element.

One of the similarities found in every region under scrutiny is the trend towards – quick (USA) or gradual (EU) – deregulation, especially as a means to eliminate or reduce monopolies. National, state-owned telecoms corporations no longer determine what is on the global agenda. According to Noam (1994: 7) the dynamics of the telecommunications policy in both Europe and the USA (this is also the case for Japan) tend towards deregulation. In the USA deregulation has expanded functionally from one

TABLE 7.1 *Information society differences between the USA, Japan and Europe*

	In the USA and Japan	In Europe
Information Content	Content is national in character.	National and multicultural in character.
	Single-language market predominates in most mass communication arenas.	Multilingual market.
	Large, integrated media companies exist with huge investment power for new technologies.	Smaller national media producers can combine forces to approach the size and financial strength of single American media producers.
	Content sellers are mostly selling to a large single market.	Sales of information products are to national and European-wide markets.
Network structures	Manufacturers and users have access to basically a single set of standards, which have evolved in place over many years. Nevertheless there are interoperability problems.	Great success in mobile telephony (GSM). Multiple standards exist at many levels; many foreign suppliers and purchasers are more familiar with the US standards.
	Most technologies are "home-grown," and developed as a result of direct national R&D programmes	Often competing technologies exist, developed as a result of national R&D initiatives; innovation is widespread.
	Wide distribution of basic infrastructure exists such as telephone, cable TV, inexpensive high-speed networks. This is less true in Japan.	Distribution of basic service is complicated by national differences in both coverage and regulatory issues. Modern telephone basic infrastructures.
Applications and Software	In the USA, a large, aggressive software product industry rules in critical basic software areas.	Specialist developers; many excellent niche market applications, e.g., in areas such as virtual reality systems, compression technologies, etc.
	In Japan, the software product industry is weak.	
The 'People' Element	Users, especially in business and industry, are technology-oriented and are reaching a level of technology maturity which increases their awareness of information use possibilities.	The historically lower level of penetration of ITC applications in business and industry, in the home and in secondary and advanced education, is now being rapidly remedied.
	Information producers, whether entertainment media, educational material, or simply personal communications, are familiar and at ease with the various technologies and applications.	Relatively late arrival of many information services to Europe means users and producers have not reached the level of their counterparts in the USA, yet Europe has some of the largest publishing groups in the world.
	Awareness of information technology and of the information society is becoming much more widespread.	Awareness activity is increasing considerably.

business sector to another. Deregulation has also spread geographically from the USA to the UK, Japan, and now to some parts of continental Europe. One of the major reasons why it took the European Commission so much longer than the USA to encourage this deregulation movement (CEC, 1987) was the huge difference in geopolitical contexts. While the USA separated network regulation from network operation as early as 1934, with the Communications Act, and the establishment of the Federal Communications Commission (FCC) as its regulatory body, many European countries were still in the process of creating their national telecommunications ministries.

Given this pro-deregulation climate, two major policy tools remain available to governments:

1 The government can act as a provider of sources of capital for infrastructure. This is a thorny issue since all governments, given their current priorities, are currently faced with very tight budgets. How to finance the necessary telecommunications infrastructure remains highly problematic. Partnerships with the private sector are and are likely to remain the most popular solution.
2 The government as a regulatory body – e.g. via the FCC, National Telecommunications and Information Administration (NTIA) in USA and European Telecommunications Standards Institute (ETS) in Europe – is involved in standard-making. Since this is necessarily based on consensus within cumbersome organizations, standard-making has proven an extremely long-winded endeavour.

Beside standard-making, US government interventions at the content level mostly focus on censorship with regard to (1) pornographic or indecent data on the Internet (Decency Act, 1995) and (2) violence (Telecommunications Act, 1996). While doing all of these things, the European Commission also looks at European content issues and the protection of Europe's cultural identity in connection with copyright and neighbouring rights (CCE, 1995). The Canadian government has an even more extensive brief, including the issue of cultural sovereignty.

The US industrial sector can rely on the highly visible support of President Clinton and Vice-President Gore when it comes to the extension of the American information superhighway. In September 1993 the National Information Infrastructure (NII) initiative was launched by the Clinton administration. The development of a broadband information infrastructure in the US will be financed by a partnership with the private sector. On 21 March 1994, on the occasion of a speech to the International Telecommunications Union (ITU) in Buenos Aires, Gore outlined the five principles according to which the United States National Information Infrastructure is to be built:

First, encourage private investment;
Second, promote competition;
Third, create a flexible regulatory framework that can keep pace with rapid technological and market changes;

Fourth, provide open access to the network for all information providers; and
Fifth, ensure universal service. (Gore in Neuman, 1994: 3)

There is nothing particularly new about four of these five conditions. Business enterprise in the USA is based on classic free market models, and the one distinguishing factor here is that they are being applied to cyberspace. Such principles as 'as few regulations and market corrections as possible' and 'innovation springs out of competition in the first place' are also applied to the telecommunications issue. Nevertheless, some functions of broadcasting, cable and other new media technologies are not like many other so-called free market industries and should therefore be subject to some form of regulation. The reasons for this are diverse, but paramount among them are the related issues of public interest and local monopoly.

With respect to the fifth condition, the Telecommunications Act (1996) officially sets out to promote competition and reduce regulation in higher-quality services for American telecommunications consumers as well as to encourage the rapid deployment of new telecommunications technologies and advanced, quality services, accessible to the largest possible user groups (including rural and high-cost areas, schools, health care service, libraries, etc.) at low rates.

As regards standard-making processes, the USA and Europe pursued (and to some extent still do) very different goals with their respective regulatory policies. The contexts in which these policy measures took shape were also very different. In 1984 the United States could be described as a fully harmonized fixed network; the FCC's priority concern around that time became to maximize the use of radio equipment by increasing inter- and intra-system competition. The reasoning behind this shift was the fact that a rapid establishment of technologically advanced and spectrum-efficient equipment could best be achieved by imposing only those few, strictly defined standards that were absolutely necessary; these should be mandated by a federal body or authority. The FCC – which adopts a participatory style – now plays a rather passive role in so far as it performs no long-term technology and/or spectrum planning. The FCC now favours a more flexible spectrum allocation policy in order to foster inter- and intra-wireless system competition.

The European Union also considers that the information superhighway is going to be a crucial tool for the continuation of political and economic integration. As early as 1987 a Green Paper outlined Europe's priority concerns in a variety of fields such as distance learning, health care, other services for public welfare, advanced services and networks in less favoured regions (CEC, 1987). Coordination regarding the future development of telecommunications throughout the European Union was set to be carried out by means of common infrastructure projects on the one hand, and pre-competitive R&D on the other, through programmes such as RACE, ESPRIT and STAR. Network development is considered crucial, entailing consideration of ISDN, digital mobile communications and the development of future broadband communications, as well as a more efficient, comprehensive, and long-term European spectrum management policy. Since equal access for all market participants is seen as another priority, the creation of a compatible European-Union-wide market for terminals and equipment as well as the establishment of open and interoperable standards are considered key requirements.

The European market, however, remains sharply different from that of the USA. Europe has traditionally been a fragmented market without the benefits of economies of scale. Consequently, users had to buy expensive terminal equipment and competition was almost non-existent for both fixed and mobile communications services. This is why, since 1984. EU regulatory policies have been geared towards two goals: (1) developing pan-European communications standards (leading to the establishment of ETSI) and (2) the introduction of competition for the provision of services. Contrary to the United States and the United Kingdom, where a clearly different pro-competition attitude was adopted earlier (in favour of free market access and/or spectrum auctioning, among other things), most regulators in continental Europe originally favoured 'limited' competition (Paetsch, 1993: 348).

One major drawback in Europe's information and telecommunications situation is certainly its so-called 'software débâcle'. Europe does not produce nearly enough mainstream software applications. With SAP – the market leader in applications software for business computer networks – a happy exception, Europe buys nearly as much software as America, but produces only a fifth as much (*Economist*, 1994: 71). In fact, things are even worse: European companies account for less than a third of sales, including in their own markets, while American software producers control over 60 per cent of the European market. 'As a result, Europe's trade deficit in packaged software is running at $18 billion a year. And things are getting worse. American firms now account for 19 of the top 30 money-makers in Europe, up from nine just five years ago' (*Economist*, 1994: 71). Among Europe's major difficulties are certainly its scattered markets, its language diversity, its different legal systems, and its different cultures.

In short, Europe's problem is basically a content problem. There is a marked similarity here with Europe's media policy which still focuses far too much on hardware development and distribution. There is a clear and incomprehensible disregard for the production of software (content). On balance, Europe's media policy is rather inadequate, and this has a decidedly negative impact on European culture and on the economic health of the European audiovisual sector. Consumers and program makers are being overlooked in favour of financial interest groups. The Television without Frontiers directive (1989) provides for the free flow of audiovisual products. However, this free flow is not backed by any financially balanced policy initiative. While the MEDIA program provides a unique counterweight to the glut of one-sided hardware development and distribution initiatives, remedying the current imbalance between production and distribution seems all but impossible.

Other countries, such as Japan, view the information superhighway as a radical solution to the growing problems of urbanization and pollution. The Nippon Telegraph & Telephone Corporation has already announced that by the year 2015 all schools, homes and offices will be wired up and interconnected through fibre optic cables. In order to prepare this large-scale project the Japanese Ministry of Post and Telecommunications initiated a three-year project (worth US$50 million) in the spring of 1994 in a bid to assess the feasibility of integrated telecommunications and broadcasting using fibre optics. Three hundred homes and offices are involved in the project. Video-on-demand, high-definition television, videoconferencing, teleshopping and telemedicine are the applications to be evaluated. Finally, the Japanese committee of telecoms experts in charge of the project considers the information superhighway as a potential means to disseminate Japanese culture worldwide. In this respect – the question of content and sovereignty – the Canadian experience may prove highly interesting.

Canada's Perspective on the Information Superhighway: Infusing some Soul into the Network

In an attempt to assess the efficacy of Canada's governmental policy on this matter, we will now examine the answers given by the Canadian government to two specific challenges: the technical evolution (the infrastructure, the hardware) on the one hand, and the cultural policy (the software, the content) on the other hand.

With regard to infrastructure, thanks to active government support, Canada can rely on an extraordinarily high telecompetitivity index (Sirois and Forget, 1995). Since the invention of telephony (in 1876), Canada has always considered the development of communications networks (always in the hands or under control of Canadians) as a priority. Other examples are the trans-Canadian telephone networks established in 1932, the Canadian Broadcasting Corporation (public broadcasting service), created in 1936, the transcontinental microwave networks built at the end of the 1950s, the launching of the first communication satellite in 1972, the move into wireless telephony at the end of the 1980s and the installation of fibre optic cables from coast to coast in recent years. And now Canadians are setting up freenets – community initiatives that provide local citizens with Internet access at attractive rates. Networks for distance learning are operational in almost every province. Schoolnet – a joint initiative between governments (federal and provincial), actors in education (teachers, universities, community colleges and schools) and industry partners – is intended to establish an electronic link between all 16,000 Canadian schools. The Canadian Network for the Advancement of Research, Industry and Education (CANARIE), an originally relatively slow-speed national network that was upgraded several times, provides the funding needed to optimize CA*net, the Canadian component of the Internet. At the same time, in Labrador and Newfoundland, efforts are being made to promote a more effective and efficient use of advanced information technology by local small- and medium-sized companies. In the autumn of 1995 a home communication service was installed in Quebec (UBI), providing interactive services such as telebanking, teleshopping, interactive mail, video-on-demand and government services on the television screen (instead of a computer screen) and operated by means of an advanced remote control (instead of a keyboard). These examples show that Canada is able to build an export-oriented infrastructure that can be considered a proactive, efficient and competitive approach in view of likely further developments on the information superhighway.

More problematic is the question of the future role of the government with regard to the development of the information superhighway. Indeed, almost everywhere (things are no different in Canada) governments have been struggling to curtail huge deficits and in many countries telecommunications monopolies are being pared down or altogether eliminated. In other words, governments everywhere no longer see themselves as being able to play the key role they used to and now rely a lot more on the private sector to finance developments.

The current pro-competition attitude *vis-à-vis* the telecommunications industry contrasts sharply with the traditionally protective reflexes concerning the production and distribution of all kinds of software products. The Canadian experience in this field is exceedingly interesting: its officially bilingual cultural market is undoubtedly one of the world's most open – and therefore one of the most vulnerable – markets to US influence. That is why the Canadian government has long been active in the promotion of Canadian content through electronic mass media and the film industry (Broadcasting Acts, 1968 and 1991). Institutions such as the National Film Board and the Canadian Broadcasting Corporation were created as part of this policy. Complementary to this is the financial support provided by the government through Telefilm Canada and fiscal compensations for investment in films shot in Canada. Moreover, legislation exists which guarantees some Canadian content on the television screens. This policy remains viable primarily through the granting and/or extension of TV and radio broadcasting licences. In practice the CRTC (Canadian Radio Telecommunications Commission) requires that 60 per cent of the programmes transmitted on public service channels (CBC/Radio-Canada) be of Canadian origin (see CRTC, 1984; Taras, 1991). Private stations are allowed to reduce this percentage by 10 per cent in prime-time (between 6 p.m. and midnight). Canadian content rules also apply in principle to pay and thematic channels. Nevertheless, Much Music (Canada's answer to MTV) and the Sports Network, two Canadian-owned, special-interest channels available on cable, respectively broadcast a mere 10 per cent and 18 per cent Canadian content (Raboy, 1990).

In accordance with its chosen cultural policy, it is the Canadian government's objective – as indicated by government documents dealing with this issue – to combat the possibility of US cultural hegemony or cultural monopoly on the information superhighway. Until now, a whole variety of governmental measures – financial compensations in film production and quotas in radio and television – have contributed to the protection of Canada's cultural identity. These policy options, together with active support of local, creative production, proved quite effective. The arrival of the information superhighway, however, is considered a potential threat. An initial, probably impulsive, response could then be to clutch at the protective measures taken in the audiovisual sector and amplify them. Precisely because of the open character of the information superhighway (homepages and websites can be designed to suit everyone's needs and tastes), the question remains whether the options chosen by the Canadian government to protect Canadian culture (mostly in the field of radio and TV) will prove applicable and effective with regard to the most recent challenge of the information superhighway.

Moreover, since Canada's weaknesses are mostly geographic and demographic, the World Wide Web is being given special attention by the government in its attempts to reverse the already pervasive Americanization of Canadian cultural products. But what is there to be done about the sheer size of its territory and the poor accessibility of remote communities in the Northern Territories, as well as the low population density – less than 30 million inhabitants heavily concentrated on a narrow strip of land along the border with the United States (whose population totals some 250 million people)? Precisely because of this, the Canadian government does not feel comfortable with policy regimes and options that are being invented elsewhere (read the USA), in utterly different geopolitical contexts. This is the reason why the Canadian government has considered it necessary to come up with policy options of its own, an approach that still favours openness and a pro-competition regime, but based on the following two tenets:

1 Canadian end-users must be given the possibility to express their own identity in order to be able to 'recognize themselves', as it were, on the information superhighway.
2 Canadians must be actively involved in the development of the information superhighway not only as consumers, but as active creators, providers of services and products, and carriers of content.

Sirois and Forget (1995: 5) spell it out as follows:

> The goal is not to keep American-generated content out of Canada but to ensure that the wonders of technology are effectively used to expand our horizons, making the full spectrum of human interest, outlooks and creativity accessible to all, with as little bias as possible, in an open environment – while at the same time creating wealth and prosperity for Canadians. There should be room for the

expression of Canadian talent and the exercise of Canadian entrepreneurship not just in Canada but in the United States and elsewhere. The question is: paying due attention to economic realities and the relatively small size of Canada and its proximity to the United States, how can such an open environment be fostered?

In Canada, the now discontinued body in charge of assessing the threats and promises of the information superhighway is the Information Highway Advisory Council. It comprises 29 experts from industry (telephony, telecommunications, data traffic, cable distribution), the cultural sector, academia and consumers' associations. This committee, whose duty it is to advise the government, was established in April 1994. In order to enlarge the knowledge base of the advisory council, it was augmented later on by another 26 Canadian experts. Priority is given to the use of the Canadian component of the information superhighway with a view to supporting local cultural and other content-related products and services. The advisory council considers that the goal must be the development of an information network of the highest possible quality at the lowest possible rates, including access for all Canadians to services that have to do with employment, education, investments, leisure, health care and overall improvement of the standard of living. Council activities are based on the following three strategic principles:

1 creating jobs by means of innovation and investments in Canada;
2 strengthening Canadian sovereignty and cultural identity;
3 ensuring universal access at an affordable price.

These principles are to be translated into policy options based on the following five working methods:

1 development of a network linked to and interoperable with other networks;
2 collaboration between government and private sector;
3 competition in the fields of possibility, products and services;
4 privacy and network protection;
5 permanent education (lifelong learning) as a key principle for Canada's information superhighway.

Competition and Regulation: A Difficult Balance

The Canadian advisory council believes that competition, rather than regulation, can and should be the driving force behind the development of the information superhighway, together with new information and communications services. Nevertheless, the advisory council unanimously recognizes the need for a national regulatory body. Again, according to this advisory council, a new role for the regulator should be urgently considered. Should it be an enforcer or an arbitrator? Central suppliers of network infrastructure, i.e. telecommunications and cable television industries, were regulated since their very beginnings. Fast technological growth and the free trade political context led to deregulation and a greater dependency upon market principles. What about the future?

The CRTC's decision, made in the autumn of 1994, in accordance with the recommendations of the advisory council, to increase the competition level within the telecommunications market, and the call for public hearings from the government with regard to the issue of convergence of the telecommunications and broadcasting sectors, are welcome responses to the recommendations of the advisory council. They clearly show that the Canadian government is taking the initiative in order to create the most favourable conditions for the establishment of new information and communication services, ensuring that the necessary infrastructure is in place. However, the precise relationship between state and the private sector remains yet to be firmly established in this vitally important area for both economics and culture.

Canadian Content

The advisory council believes that government policy must aim to enhance Canadian identity while promoting the development of the information superhighway. The council thinks that any recommendation made with respect to the information superhighway should be based on the provisions of the 1968 and 1991 Broadcasting Acts, especially since the issue of Canadian content is intimately related to employment matters. The advisory council recommends that exports of Canadian television programmes and film productions should be facilitated in the future. Harmonization of national and provincial support funds is therefore necessary. The council makes it clear that the principle of freedom of speech as enshrined in the Canadian Constitution must not be undermined on the information superhighway.

Similar to what has been happening in the other regions under scrutiny, a Canadian Working Group on Access and Social Impacts has examined ways to reformulate Canadian legislation with regard to offensive content transmitted through new information technologies. Since one can already find plenty of pornographic, obscene and hateful material in cyberspace (including Internet newsgroups), and given the fact that such content is easier to disseminate in digital format and therefore difficult to control, the working group will provide the Canadian legislator with recommendations for amending parts of applicable legislation, so as to more efficiently control children's access to such content.

Conclusion

Converging technologies are blurring the boundaries between telecommunications, broadcasting, computing and data information services. Such services as we now know them will be things of the past. This chapter has shown among other things that governments are faced with confusing and sometimes downright chaotic situations. It all boils down to fighting vertical and horizontal alliances of content and service providers and manufacturers. National and supranational governing bodies are scrambling – rather ineffectually – to regain some control over the essentially horizontal, cross-national connecting abilities of the information superhighway and other advanced telecommunications systems. It must be clear, however, that governments still have a role to play. Above all they must strive to resolve the infrastructure situation, which is currently characterized by the emergence of mega-mergers of suppliers. Furthermore, governments must ensure that power, currently in the hands of a few, is better distributed among larger groups of actors. Otherwise the only force in operation will be the market, which means that too many people may be left behind once and for all.

We saw that there is at least one government that is determined to do something about this state of affairs. The Canadian government is taking steps to prevent the information superhighway from becoming a vehicle for cultural homogenization or an outlet for monopolies. Canadian policy-makers want to make sure that their opinions on content issues are heard on the international scene, which means that Canadians may become more actively involved in the global coordination of the information superhighway. A lot of other regions, including the European Union, have also been looking for an approach which is more strategic and user-friendly. And most of them agree that the USA is never more aggressive in international matters than when it sees in them a means to boost the US economy.

Regardless of future policy options, there is a clear need for a complete change of paradigm concerning the regulation of infrastructure use and the production and selection of content on the information superhighway. The unavoidable international dimension of this global competition game must be reflected in policy measures. Advancing globalization will also prove useful in the long run inasmuch as it should help eliminate many of the multiple variants of existing standards. Based upon a realistic assessment of the needs of the global marketplace, it must be clear that truly international standards are becoming increasingly necessary. Moreover, different user groups could play a vital and constructive role in the establishment of truly useful standards. Therefore, not only major user groups (multinationals or public administrations) – which already sit at the various negotiating tables – but also professional or trade associations and individuals using micro-computers should have their say, since they 'share common expectations about standardization' (Ferné, 1995: 457). A number of national governments and supranational bodies such as the European Union are determined not to let government regulations or isolation from the rest of the world become obstacles to the development of the information superhighway, in which openness and interoperability remain crucial. What really matters is to use technology to recognize and stimulate creative activities, so that anybody who wishes to produce and distribute new products and services (from electric cars to electronic music) may do so, to make institutions more flexible, to eliminate market boundaries and expand horizons, and perhaps ultimately to help draw humankind together. Such a goal, however, requires a policy mix of privatization, regulation and subvention to ensure that open standards do not result in international monopolies.

References

Bouwman, H. (1994) 'De doos van Pandora', in H. Bouwman and S. Pröpper (eds), *Multimedia tussen hope en hype*. Amsterdam: Otto Cramwinckel.

CCE: Commission des Communautés Européennes (1995) *Livre vert. Le droit d'auteur et les droits voisins dans la Société de l'Information* (COM(95) 382), 19 July. Brussels: CCE.

CEC: Commission of the European Communities (1987) *Towards a Dynamic European Economy. Green Paper on the Development of the Common Market for Telecommunications Services and Equipment* (COM(87) 290) 30 June. Brussels: CEC.

CRTC: Canadian Radio-Television and Telecommunications Commission (1984) *Public Notice 1984–94 on Recognition for Canadian Programs*, 15 August. Ottawa: CRTC.

d'Haenens, L. (ed.) (1998) *Images of Canadianness. Visions on Canada's Politics, Culture, Economics*. Ottawa: Ottawa University Press.

Economist (1994) 'Europe's software débâcle', 12 November: 71–2.

Ferné, G. (1995) 'Information technology standardization and users. International challenges move the process forward', in B. Kahin and J. Abbate (eds), *Standards Policy for Information Infrastructure*. Cambridge, MA: MIT Press.

Lehr, W. (1995) 'Compatibility standards and interoperability: Lessons from the Internet', in B. Kahin and J. Abbate (eds), *Standards Policy for Information Infrastructure*. Cambridge, MA: MIT Press.

Libicki, M.C. (1995) 'Standards: the rough road to the common byte', in B. Kahin and J. Abbate (eds), *Standards Policy for Information Infrastructure*. Cambridge, MA: MIT Press.

Longhorn, R. (1995) 'The information society. Comparisons in the trio of Europe, North America and Japan', *I&T Magazine*, 16: 5–9.

Neuman, W.R. (ed.) (1994) *Toward a Global Information Infrastructure*. Washington, DC: United States Information Agency.

Noam, E.M. (1994) 'Is telecommunications deregulation an expansionary process?' in E.M. Noam and G. Pogorel (eds), *Asymmetric Deregulation. The Dynamics of Telecommunications Policy in Europe and the United States*. Norwood, NJ: Ablex.

Paetsch, M. (1993) *Mobile Communications in the United States and Europe: Regulation, Technology, and Markets*. Boston and London: Artech House.

Raboy, M. (1990) *Missed Opportunities: The Story of Canada's Broadcasting Policy*. Montreal: McGill-Queen's University Press.

Sirois, C. and Forget, C. (1995) *The Medium and the Muse. Culture, Telecommunications and the Information Highway*. Montreal: Institute for Research on Public Policy (IRPP).

Taras, D. (1991) 'The new undefended border. American television, Canadian audiences and the Canadian broadcasting system', in R. Kroes (ed.), *Within the United States' Orbit: Small Cultures vis-à-vis the United States*. Amsterdam: VU University Press.

8 Technocities and Development: Images of Inferno and Utopia

Simon Bell

What Is Urban Migration?

The core interest for this chapter is the juxtaposition of the massive and rapid development of global communications and the related development of social structures which follow from the population of what has been variously called the 'Technocity', 'the invisible city', 'Teletopia', 'the information city', 'cyberspace', etc. (see Graham and Marvin, 1996: 9) set against the context of the developing world. A major point which I will begin and end with is that the development of any urban morphology is not a painless or uncontentious issue – but it does provide rich learning potential for those who wish to learn (the lack of attention paid to this issue by serious scholarship is a point also made and developed by Graham and Marvin, 1996: xiii–xiv).

This chapter does not profess to be definitive or unduly authoritative. Rather it represents an exercise in self-analysis and a consideration and reappraisal of ten years' experience in development-related research and consultancy. Following from this, experience is set against the context of the 'technocity' phenomenon.

First, a little nostalgic background. When I was studying international development as an undergraduate in the late 1970s and early 1980s, one stock theory at the time was that of urban bias based upon the evidence of mounting urban migration. The faculty of the university where I studied and elsewhere were largely (though not exclusively) interested in rural development issues (such as land and water development and conservation, ecology issues, agricultural project management, facilities provision for rural populations and healthcare) and global economics. The rising power of the urban conurbations in developing countries appeared to be seen by some faculty as being both beyond their scope and also as unattractive (for example, see the weakness of this area in Livingstone, 1982).

I remember a sense of puzzlement shared by myself and others in the student body at the comparative lack of attention which the cities of the Third World (as we referred to the developing nations at the time) received. A view held by myself and some of my more roguish colleagues was that rural development was a rather more attractive prospect for university lecturers and students than its urban counterpart. Rural development could be related to interesting and exotic cultures, wide African landscapes, rainforests, wildlife and exotic flora. Link this phenomenon to a philanthropic interest in the ecological issues of the day and it is easy to see why academics might find the rural concept appealing. Urban development with its shanty towns, squalor and disease might not have seemed to have so much to recommend it. Urban landscapes are certainly not attractive in the development context.

Books like that of Paul Harrison (1984) were informative for those who wished to follow the urban development argument. Harrison, in the section 'Hell is a City', sets out the case for the urban tendency in development. A brief description of some of his major points might strike some familiar chords with those now looking out on the new cyber cityscape.

The modern urban city, according to Harrison, was set up by the colonial elite and developed as the living quarters of those arriving in developing countries to expropriate wealth. These colonial settlers were interested in 'providing themselves, in their city bases, with the services and salaries they were used to back home' (1984: 148). But since the development of these colonial 'islands of privilege' the cities of the developing world have expanded beyond all concept of planning or order: 'In 1940, Third World towns and cities housed 185 million people. By 1975, the number had risen to 770 million' (1984: 145).

The rural areas' population also grew but the development of cities far outstripped population in the countryside. No definitive argument can be produced for the expansion of the cities. Rural push, urban pull, lack of family planning in the cities – whatever the reason the population of cities expanded dramatically. Why? Harrison's argument relates mainly to economic factors. Cities had the lion's share of resources set aside for development. They had more jobs, more health centres and schools. The rural areas rarely provided children with the necessary resources to rise above the grind of bare subsistence. In the minds of the people of the developing world, cities were best. Apart from anything else they offered a potential gateway to the wealth and advantages of the richer nations. But cities, acting as magnets for resources, also acted as gateways through which wealth could be exported. This long-term export could lead to another quality of urban development – mass exploitation. Gandhi argues:

> The British have exploited India through its cities, the latter had exploited the villages. The blood of the villages is the cement with which the edifice of the cities is built. (quoted in Harrison, 1984: 137)

To bring this debate into the present context: are there current parallels between this view of the historic development process of cities in the Third

World and the pull of the Internet? Do we have modern electronic equivalents to traditional problems with urban life? Cyber centres of privilege for those who have the right to migrate? Urban sprawl in cyberspace? What of the development of the electronic shanty? Is the current fixation on the value of the all-encompassing Net influencing a new generation of economic and social migrants to conform to the apparent uniform, Western culture of the electronic frontier?

Before looking again at the potential for the technocities to impact upon populations in developing countries, it is worth looking in a little more detail at some of the theoretical and empirical background.

A Fractured Debate – World Realities and Technological Realities

The Internet / IT, communications specialists and the technologists, on the one hand, and the researchers and analysts in development studies, on the other, show evidence of a wide variety of views and ideologies. The two fields differ from each other in terms of their worldviews at a generalized level as well as being riven within themselves between various factions (see, for example, Hettne, 1990 and So, 1990).

In addition, we have the existing discussions within the development community. This discussion is generally seen as being pessimistic about the development of world history, about the present condition of global humanity and often about the potential for the future. As an undergraduate I was continually shocked by the gloomy forecasts and the depressing possibilities which the world faced (for an example of the material which confronted us see the ecology, resources and population discussion in Ehrlich et al., 1977).

Set against development studies, we have the body of literature derived largely from the technologists, which supports the view that technology can be a major spur to development *per se* (e.g. see Woherem, 1993). The technologists are usually optimistic enough. All things are and will be possible. Technology will answer all human problems, from feeding the growing world population by miracle yields to controlling population growth itself though the removal of want. According to this community the growth of global communications is a great opportunity. Views concerning the value of technology in the process of development, however, vary from the opportunism of Woherem (1993) to the studied analysis and empiricism of Lane (1990) and the bleak forecast of Litherland (1995). Lane, in particular, draws attention to several case studies indicating the value of Internet communications in providing cheap and reliable forms of national and international communications. The value of this to agencies seeking to gain funding from the wealthy nations and wishing to be heard in international forums is unquestionable.

The contrast between theorists and practitioners in development studies and technologists working in the developing countries can appear stark. Despite a great deal of communality between individuals, the overall impression remains that there are 'two camps'.

The camps can be set in the context of a traditional debate in development theory – on the one hand, dependency theory lives on in the guise of the critical view of technology fixes, usually seen as unsustainable in context. The technology which can provide information also re-emphasizes dominant flows of economic and political control which feeds into local elites but which has little effect upon the overall development of nations. On the other hand, the 'modernization' school, lamented in the late 1970s by some but never too far away, is evident in much of what passes in debates among the technologists. Modernity can provide the developing nations with a means to 'leap-frog' into the information age.

The 'fracture' indicated in the title of this section relates to the contrast between the two views of the technology increasingly being applied globally. In the next section the prospects for technocities in the developing world is reviewed.

The Development Context – Some Views from the Periphery

This chapter does not claim to represent scientifically sampled, statistically significant views of the adoption of global networking. Rather it is a learning cycle review arising from the research and consultancy experiences of the author over 10 years.

The major issues open for review here are as follows:

Is the technocity understood locally (by whatever label)?
Do the people in context know what it is?
Are there perceived benefits of use of/ migration to the city?
   This is a tricky question. The main thrust is to discover if the perception of the use of the technology is as a valuable thing for specific reasons or if it is seen as being a good thing in itself irrespective of what an organization might do with it.
What does this migration mean to the process of development?
   The most difficult question of all. In asking what this means to development I am asking if the technocity is seen as providing some initiative or insight which will provide a deep development benefit (economic or other).

Before going on I would set the scene concerning my qualifications to embark on this exercise. This requires a little self-analysis and a view of the learning cycle relating to the position I have arrived at.

Now Where Was I?

I have been working on information technology (IT) and information systems (IS) related projects in developing countries since 1986. Generally I have worked with public sector agencies in the development of IS projects and have been centrally concerned with methodology development (Bell and Wood-Harper, 1998). My work has covered some of the countries of Asia as well as East and West Africa. Overall, the result of my experience tends towards the pessimism of the development studies camp rather than the optimism of some of the technologists. In the process of my own analysis I have undertaken a fair degree of self-reflection on the results of experience (Bell, 1996). This has inclined me to adopt a view that all consultancy and research should involve a consciously applied method for self-learning or a 'learning cycle', whereby experience can be sequentially fed back into consideration of subsequent action. This should occur so that action can make effective use of past experience. Such reflection on praxis can refer to the individual researcher, reflection by the population involved in a project or independent reflection by both. Figure 8.1 depicts one view of the learning cycle. This is further developed on p. 162.

It is hard to compare unalike cases but in the following section I want to develop a framework for comparing a number of cases and reviewing the

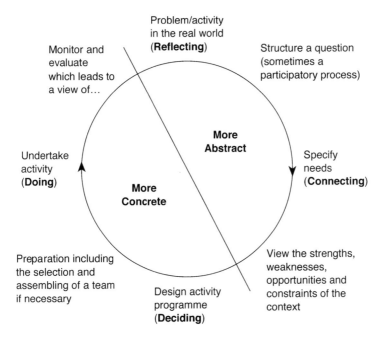

FIGURE 8.1 *A learning cycle (Adapted from Kolb (1984) and building upon Senge et al. (1994, p. 60).*

implications for the development of international communications in order to provide orders of magnitude improvements in development processes.

A Framework for Analysing Experience

The questions which this chapter sets out to address will be the criteria by which the three case studies of material are reviewed. It is not desirable here to go into detail but the case studies are drawn from Nigeria, Pakistan and China.

The three case studies, as indicated earlier, are quite significantly different in terms of the areas of the public sector they represent and the types of projects they encompassed. However, in each case the issue of communications and computers was prominent and there was considerable interest in the potential which these technologies offered.

Nigeria

In Nigeria my reflections span three projects, starting in 1986 and continuing to the present day. All three projects were concerned with education and the use of IT and two of them were directly concerned with communications.

Is the technocity understood locally (by whatever label)? The idea of global communications has been gradually taking shape in the observations and requests of Nigerian projects with which I have worked. There is a perception that the Internet is a global, cheap form of communication and that it is, therefore, a good idea for developing nations to get access to this. The reasons for this are less easy to discover.

Are there perceived benefits of use of/ migration to the city? My work in Nigeria was largely with the academic community. Among this group access was mainly, if not exclusively, seen as a means to make contact with the global community and to gain cheap access to international resources of knowledge. There was much vagueness as to how it worked and the specific details of cost.

What does this migration mean to the process of development? The phrase that probably sums up my experience is 'globalization'. In this is included the perceived need by local academics to be communicating with a global academic community, the need to be 'relevant' and produce work of importance and the fear of continuing to lose ground to the Western scientific community.

Pakistan

In this context I was working for the United Nations (UN) in a remote rural area in 1995. My reflections here concern the views of local administrators both in the Pakistan government and the UN. The issue of communications and IT is seen by some of the local stakeholders as being of central concern.

Is the technocity understood locally (by whatever label)? The capacities of the Internet were known in part at least and the desire for a communications device for sharing information nationally and internationally was appreciated by a growing cross-section of the professionals I worked with. However, there was a general lack of informed opinion of the potential which the Net offers and the down-side (e.g. the perils of over-reliance on non-sustainable technology) and costs of access.

Are there perceived benefits of use of/ migration to the city? The main feature of this aspect can be set out as low cost and reliable communications specifically relating to assisting in the coordination of local projects. In this sense the technology is seen as a means to develop sophisticated regional, national and international communication devices.

What does this migration mean to the process of development? The primary concern relating to capacity for greater communication is linked here to the secondary concerns for increased control, accountability and improved management. The only view of 'city' or community within this is the project community or, maybe, the UN community.

China

In China my work was with an agency which monitored national and international economics and was concerned with national policy on pricing of commodities. The work took place over an 18-month period in 1993–4.

Is the technocity understood locally (by whatever label)? Immense local interest was focused most specifically on the potential which electronic communications offered in terms of global information relating to exchange rates and commodity prices.

Are there perceived benefits of use of/ migration to the city? The main thrust here was the issue of access to international information and the capacity to download this and use it as a basis for local modelling exercises. Capacity to influence and be influenced by global information was a secondary, although important issue.

What does this migration mean to the process of development? Here the focus was primarily upon improved knowledge, improved speed of reaction to changes in volatile markets and improved control over prices. There was no real perception of the technology embracing the concept of community.

Urban Migration and Internet Migration – Is There a Link in the Morphologies of These Two Phenomena in the Developing Country Context?

The single, overarching observation arising from the three case studies is the perception of the value of and existence of global communications on-line. Individuals and agencies in the developing world want to be part of the discourse and, for a variety of reasons, expect direct benefit from it. Perhaps the most surprising observation arising from the case studies is that at the time of writing none of the agencies described is actually using or has access to the Internet directly. The main reasons appear to be:

- difficulties in gaining the technical access (includes problems with local telecommunications);
- problems with the politics of access. This includes dealing with who has access to this form of communication and what freedom they have to communicate on potentially sensitive issues. This political dimension then feeds into the third item;
- problems with funding the link. Initial funding is often available but the budget for recurrent, year on year costs is not forthcoming.

We need to recognize therefore that in all three cases perception has yet to be matched against the outcome of experience.

The purpose of this section is to compare the urban migration view of development with the current development of on-line communities on the Internet, the technocities.

An important aside should be made at this point. The idea of communities in cyberspace called technocities is not a current or active idea among the communities mentioned in the three case studies. The interest is mainly in terms of communication, not community. It should be noted that if urban migration in the electronic sense is occurring, this activity is largely not conscious as yet. The use of the Internet, e-mail, etc. is seen purely as a means to improve communications. The prospect of this developing into an electronic community, eventually of city proportions, was never discussed and has not arisen to date.

With this key consideration in mind, I will go on to set out some of the bases for comparison. We can take Harrison's (1984) discussion and see, in generalized terms, city morphology in developing countries as containing a number of stages:

The colonial city – primarily political centres, privileged dwelling places for the colonists. Access to indigenous populations strictly limited, most particularly in the elite areas.
The magnet city – provoking mass in-migration, mainly for economic reasons;

The city as hell – overpopulated but still drawing rural populations. A place of shanty and slum.

These views of city development need to be considered in terms of the potential symptoms of urban migration to the Internet. As was stated earlier:

> Do we have modern electronic equivalents to traditional problems with urban life? Cyber centres of privilege for those who have the right to migrate? Urban sprawl in cyberspace? What of the development of the electronic shanty? Is the current fixation on the value of the all-encompassing Net influencing a new generation of economic and social migrants to conform to the apparent uniform, Western culture of the electronic frontier?

The current position, in terms of the description set out, seems to be somewhere between the colonial city and the city as magnet. There appears to be a perception across the three case studies, right or wrong, that on-line access to the Internet is a very positive opportunity, that it is a place where users have access to a wide (almost unimaginable) range of services and facilities. This is the magnet drawing academics, aid planners and public sector employees. So, in answer to the points made in the quotation above, yes we have evidence of the technocity acting as a centre of privilege, and yes this is inducing (particularly in the Nigerian case) a new generation of social and economic migrants. However, at present it is too early to comment on the potential for cyber-sprawl and shanty. Maybe these would be indicative of the technocity evolving from city as magnet to city as hell. We do not appear to be there yet but there are some worrying signs of movement in that direction. The literature is full of complaints about the difficulty of navigating in cyberspace (a labyrinth?), of congestion, of not being able to find specific people and places in the maze of directories and systems, of subversive and dark ghetto areas (see the regular gripes and groans in *Internet* magazine) or of cyber-mugging (see *Computer Weekly*, 16 November 1995: 20).

I come to no firm conclusion as to where the potential and actual users in the developing countries are in their relationship with this new urbanism, new technology. The position of the aid agencies is interesting. Grant Lewis and Samoff (1992) as well as Hanna and Boyson (1993) indicate dramatic increases in the use of IT by development agencies like the World Bank. However, it is also evident that there are considerable problems in application and a lack of agreement between donor and recipient on the direction in which IT use in developing countries should be going (see Litherland, 1995). It is too early to make definitive statements but there is a fair degree of evidence, arising even from my survey, that parallels do exist between two city morphologies. A heartening aspect is that 'city as hell' does not appear to be strongly apparent as yet, so there is still time for consideration and planning.

Learning Cycles

Looking beyond the immediate debate: have aspects of discussions about impacts of revolutionary technologies been seen before? Both the green revolution and the breaking communications revolution have claimed their place in history. Both have had major impacts but both have similarly been viewed by many impartial observers as missing the point and causing as many problems as they purport to cure.

With the advent of the technocity there is a need for a wider and more informed vision of the possibilities and limitations which this technology provides. Further, there is clear need for a mature and committed debate concerning our understanding of the ramifications of the processes of change which the technologies invoke. The current situation *vis-à-vis* the development of technocities provides a rich learning opportunity if the lessons are effectively absorbed.

I return to the learning cycle. The use of the Internet to develop a global community in cyberspace appears to have the potential to develop unhealthy correspondences to the development of the cities of the developing countries, for example as islands of privilege, resources for the few, shanties, slums, centres of disease. In considering the foregoing discussion as a learning exercise I pursued the process as is shown in Figure 8.2.

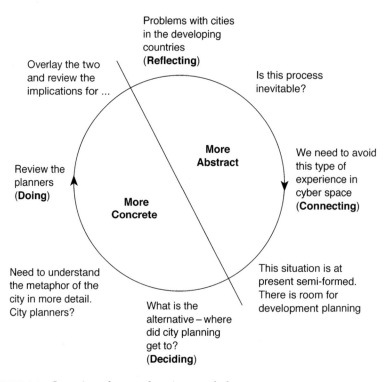

FIGURE 8.2 *Learning about technocity morphology.*

The principle arising from the cycle is the correspondence of city/urban planning with the evolution of the technocity. If there is a correspondence between the growth of cities in the developing countries and the technocities, there may also be metaphoric correspondences and lessons to be learned from the ideas of urban planning.

The discussion so far and the comparison of urban bias and technocity development has been a metaphoric comparison. To extend this into a new debate, we can set it against the following metaphors:

islands of privilege, resources for the few, shanties, slums, centres of disease;

Alternative metaphors are:

urban renewal, green belts, garden cities, centres of employment, quality dwellings, place of community action or of participatory democracy.

It is not the intention of this chapter to develop this comparison further; instead I hope to develop discussion, provoke debate and seek correspondences between the technocity debate and earlier discussions which might cast comparative light upon the subject. In the following few pages the view of city planning is considered and some interesting (though provocative) views are described.

For the purposes of reviewing city planning, I was drawn to look at the work of Patrick Geddes (1854–1932) from whom many of the terms set out in the alternative metaphors above have been borrowed (these two sets of metaphors conform to his view of cities as being either 'paleotechnic' or 'neotechnic' (Geddes, 1949: 32)). Geddes is a little-known person today outside planning circles although he is acknowledged to be the father of modern town planning (Stalley, 1972) and responsible for the analysis and investigation of city morphology as well as the supplier of modern terms like 'conurbation' and 'megalopolis'. There is not time to do justice to Geddes here other than to make the observation that he foresaw the development of the modern city and was appalled by the possibilities which growth without highly principled planning could mean to the lives of individuals. Alluding to the development of 'evolutionary cities' and describing what would happen if the work of Geddes remained relatively unknown, Stalley laments:

> In the absence of this [Geddes's ideas], planners, developers and bureaucrats will continue to operate on a 'project' basis without an understanding of the delicate interrelationships between all elements of the community, the community of mankind and the community which is the world. (1972: xiv)

This is, I feel, very relevant to the current situation in the cities of the developing world as well as to the current discussion. Geddes has been called the 'spokesman for man and the environment' (Stalley), in that he recognized that city developments affect people, the environment and the world and that they are often not developed at the level of and for the good of the individual; rather they tend to dominate and enslave. For Geddes the old idea of city was the palaeotechnic: his view was that a new concept, the neotechnic, was necessary to understand the evolution of the city:

> The transition from Paleotechnic to Neotechnic is in actual progress around us; yet in need of strongly emphasising these two types of evolution as Inferno and Eutopia respectively. (Geddes, 1949: 46)

Geddes's city as inferno relates directly to Harrison's view of 'City as Hell'. To assist in the development of the neotechnic city Geddes advocated the need for humanity to develop cities in line with a cyclic development of themselves. For this purpose he produced a diagram which he referred to as the 'Notation of Life' (Figure 8.3). This is in itself a 'learning cycle' but indicated in Geddes's terms what was needed for the city to develop.

His diagram links psychology with sociology, economics with ideation and pragmatics with mysticism. For Geddes the multidisciplinarian (referred to by Stalley as a Renaissance man) the city indeed arises from a completed development. His thinking is now dated in many aspects but much of the content is still quite relevant and provocative, specifically with regard to the

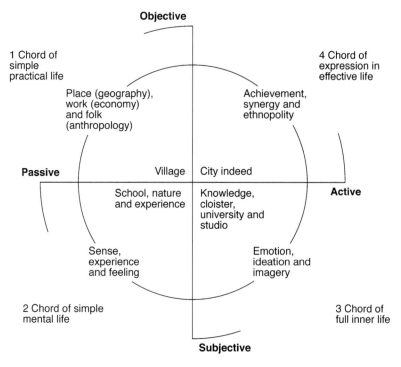

FIGURE 8.3 *Geddes's Notation of Life (Adapted from Geddes, pp. 194, 195, 201, 204)*.

view of self-development as a necessary prerequisite for social development. In the diagram, the cycle of social life or the life of the individual passes from the pragmatics of work, place and people (as basic concepts of social life) to the ideal of social synergy released and developed in the true city.

For Geddes the first stage is rooted in passivity and objectivity. It provided the requisites for the development of sociology and provided the foundation for social life in village and town. For cities to arise at this point in the cycle there would be deficiency of intellectual and spiritual content. The 'city indeed' requires this inner content of its citizens and these are only to be gained in subjective, more abstract development. The next stages were vital, being interior to society and individual. He argues that sense, experience and feeling are the subjective counterpoints in human beings to the objective experience of life. However, these elements of what he calls 'simple mental life' are developed further and supplemented in the third stage – that of emotion, ideation and imagery. Although still subjective, Geddes argued that the development of knowledge and mysticism at the third stage provides society with the possibility of a more perfect, active social structure than that which was possible in the first stage. Through these levels of inner development, Geddes proposes the foundation of the true city, enhancing and building upon the effective aspects of the earlier stages.

Central to his argument is the notion that cities cannot develop effectively until people have the necessary synergy to provide sound foundations. This was not the end of the process. Cycle will follow cycle, developing in itself and extending into a spiral as one cycle is completed (Figure 8.4).

The value of the 'Notation' to the discussion of the development of the technocity is as a cautionary note. In Geddes's view there was an integral parallelism between the development of the individual, the society and the related development of the city morphology. Cities could and would become hell if each was developed in isolation.

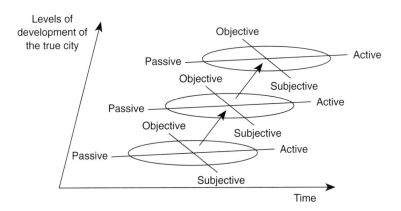

FIGURE 8.4 *A learning spiral based on Geddes's Notation of Life.*

The interesting point here is to reflect on the current development of the technocity. What is the order of the process of evolution? Are we making progress towards a cyber-inferno or even a cyber-utopia?

As the technocity develops and extends into more and more houses and offices and as the people of the developing world demand access, the concept of a global community takes shape in a new guise. It seems that at present this development is either under the hand of single enterprises like Microsoft or developing without plan or morality as the Internet has done to date. As more and more lives are touched by the technocity, and as its power continues to wax it may be that the means to avoid a depressing descent into the dark side of such a community (as seen in the 'Paleotechnic' of Geddes's work) is for Bill Gates and the other self-selected technocity planners to learn from the work of those who grappled with the same problems but in a different context.

But this outcome is only half of the equation for mature social control of technology. This outcome would be unduly technologically deterministic, in which society is moulded and structured at the whim of the controllers of technology. The real learning within the technocity would be seen as having reached maturity if it were seen to be the responsibility of the users of the city, the citizens of this new space, to take a proactive hand in ensuring that the technocities are places designed for as well as by society, i.e. that they offer an acceptable cultural, social and political environment. This is another potential arising from the disparate and uncontrolled nature of the Internet. The development of this global, empowered citizenry, actively developing the structures and contents of the technocity, would itself be evidence that the technocity is developing in a manner which Geddes would perceive as positive, in that individual development was being mirrored in social structures to which the people of the less developed countries could contribute.

There is a great need for informative studies which are focused on reviewing such social action within the advancing social colonization of the technocity. Future work needs to address the problem of seeing technological determinism as the primary driving force. A more profitable basis for analysis is to start from the rubric that technology is part of society. From this point our focus of study shifts to reviewing how 'society influences technology' (Graham and Marvin, 1996: 105).

References

Bell, S. (1996) *Learning with Information Systems: Learning Cycles in Information Systems Development*. London: Routledge.
Bell, S. and Wood-Harper, A.T. (1998) *Rapid Information Systems Development: Systems Analysis and Design in an Imperfect World* (2nd edn). Wokingham: McGraw-Hill.
Ehrlich, P.R., Ehrlich, A.H. and Holdren, J.P. (1977) *Ecoscience: Population, Resources, Environment*. San Francisco: Freeman.

Geddes, P. (1949) *Cities in Evolution*. London: Williams & Norgate.
Graham, S. and Marvin, S. (1996) *Telecommunications and the City*. London: Routledge.
Grant Lewis, S. and Samoff, J. (eds) (1992) *Microcomputers in African Development*. Boulder, CO: Westview.
Hanna, N. and Boyson, S. (1993) *Information Technology in World Bank Lending: Increasing the Developmental Impact*. World Bank Discussion Papers, no. 206. Washington: World Bank.
Harrison, P. (1984) *Inside the Third World*. Harmondsworth, Middlesex: Penguin.
Hettne, B. (1990) *Development Theory and the Three Worlds* (Longman Development Studies). Harlow: Longman
Kolb, D. (1984) *Experiential Learning: Experience as the Source of Learning and Development*. Englewood Cliffs, NJ: Prentice-Hall.
Lane, G. (1990) *Communications for Progress*. London: Catholic Institute for International Relations.
Litherland, S. (1995) 'Communications: growth of Internet leaves poor behind', *Information Technology in Developing Countries*, 5(4): 18–19.
Livingstone, I. (1982) *Approaches to Development Studies: Essays in Honour of Athol Mackintosh*. Aldershot: Gower.
Senge, P., Ross, R., Smith, B., Roberts, C. and Kleiner, A. (1994) *The Fifth Discipline Fieldbook: Strategies and Tools for Building a Learning Organisation*. London: Nicholas Brealey.
So, A. (1990) *Social Change and Development: Modernization, Dependency and World Systems Theories* (Sage Library of Social Research 178). London: Sage.
Stalley, M. (ed.) (1972) *Patrick Geddes: Spokesman for Man and the Environment*. New Brunswick, NJ: Rutgers University Press.
Woherem, E. (1993) *Information Technology in Africa: Challenges and Opportunities*. Nairobi: ACTS Press.

PART 4

PERSPECTIVES

9 Designs on the City: Urban Experience in the Age of Electronic Reproduction

John Pickering

> The information society is on its way. A digital revolution is triggering structural changes comparable to last century's industrial revolution, with correspondingly high economic stakes. The process cannot be stopped and will lead eventually to a knowledge-based economy.

Techno-enthusiasm like this, often accompanied by fanciful illustrations of the wonders of cyberspace, fills popular magazines and bookshops. As the next century approaches, digital technology is celebrated with the fervour aroused by heat engine technology at the close of the last one. The speed with which information can circulate, whether between people at home or on the move, within corporations and workplaces, between the governors and the governed and around the whole globe, is now central to the story of cultural progress that people in technologized societies are telling themselves. In advertising the computer is an icon of precision, power and assured worth. The arcades are digitized cornucopias from which tumble glittering cascades of electronics. Machines that not much more than a decade ago were tools for research elites are now commonplace consumer durables participating in mundane social life (Pickering, 1997). The technologizing of the human condition depicted in cyberpunk fiction seems to be coming up over the horizon faster than it was (Sterling, 1986; Haraway, 1995; Gibson and Sterling, 1992).

However, the above quotation is not fiction, nor is it from a catalogue or magazine. It is the opening section of a recent report to the European Commission entitled *Europe's Way to the Information Society: an Action Plan* (European Commission, 1994). It saw the Internet and other aspects of digital technology as: 'an acceleration of the liberalisation process and the achievement and the preservation of universal service and the Internal Market principles of free movement. . . . The deployment and financing of an information infrastructure will be primarily the responsibility of the private sector.' The tone of the report was urgent: 'The race is on at global level, notably with USA and Japan. Those countries which will adapt themselves most readily will de facto set technological standards for those who follow.' This tone is similar to that of a number of reports produced in the G7 nations over the last decade or so. The feeling they express is that digital technology will continue its rapid growth and penetration of all spheres of life, that it will help economic growth, that communication networks are its most distinctive manifestation and that these will be a positive force in maintaining democratic regulation of societies.

The rapid deployment of public and transnational corporate resources to create networks is now a central policy issue for all G7 nations. This is in many ways similar to the building of railways in the middle nineteenth century. It is seen as a multiplier that will enhance economic and cultural structures already in place. The social changes it will make to the way we live are represented as being cohesive and liberating. For example, during the early 1990s the UK Labour Party used information technology in education and industry as a symbol of its modernization. The Clinton and Gore re-election campaign used it as a sign for Democratic progessiveness in contrast to Republican conservatism. Al Gore in particular has for the past decade made it part of his political profile to urge the US Treasury to treat the information superhighway as part of the economic infrastructure, akin to the state highway system.

Digital technology now symbolizes where world culture is heading. It signifies economic progress and openness in communication and governance. Although rapid change and exaggeration make it difficult to evaluate the cultural role digital technology is playing, it is interesting to note that sensationalized accounts are sometimes modest when compared with the objectives of research corporations. Popular magazines may carry articles on virtual reality sex games, but it is in internal publications of research laboratories that we find discussions of matter transference and of electronic implants for the direct creation of pleasure (Pearson and Cochrane, 1995). Clearly those with the power to develop this technology are taking seriously the cliché that is often used about it: 'The only limitation is imagination.'

This chapter will examine some ways in which digital technology may influence urban living. The suggestion will be that it will be modest in the short term, despite the substantial changes it is making in economic and political life. In the longer term, however, there may well be a change in the way cites are inhabited and designed. The chapter will conclude by suggesting that Walter Benjamin's prescient understanding of the cultural role of technology provides a powerful framework within which to understand this change. To begin, we look briefly at the evolutionary origins of the city.

Evolution and the City

By 'evolution', most of us understand Darwin's biological theory and its subsequent elaboration into its contemporary form where genetic transmission holds central place. Like any hypothesis, the theory is a conjecture, not a fact. It cannot be considered final or complete, and recent work is changing it radically. What is emerging is a very different picture of evolutionary change in which the transmission and modification of genetic information is only part of the story (Goodwin, 1994). In particular, more attention is being paid to individual development and it is more than ever clear that culture rather than genetics underlies the emergence of modern human beings. For example, there was an exponential increase in the variety and sophistication of stone tools during the late palaeolithic period, that is, from about 100,000 to 10,000 years ago. This period covers the emergence of modern human beings and, towards its end, the appearance of the first cities. However, during this period nothing in the genetic makeup of human beings changed radically.

This confirms that it is a mistake to think that it is essentially something inside human beings that makes them different from animals. Though there are differences of course, far outweighing anything internal is the fact that in the human case, development occs in an environment that is a cultural product, the city being the most noticeable example of this. This means, in a more radical sense than hitherto considered, that human beings are 'self-produced', just because the possibilities for action, which shape development, are provided by the human environment which is the material trace of previous human action (Kingdon, 1993). What was changing during the late palaeolithic period was the cumulative system of cultural relations that supports the human condition. The human condition is the system. It has evolved components both outside and inside the human body, but no one part is crucial.

To appreciate this mutuality of biology and culture properly requires appropriate theories. A number of recent accounts of the interaction of human biology and the social context during development favour a systems view of genetics and culture (Edelman, 1992; Maturana and Varela, 1992; Johnson, 1993). For example, Edelman suggests that the brain is more fundamentally plastic, and thus more influenced by the environment, than was previously thought. Recent research into brain development has extended the sensitive period of brain growth from two to around ten years (Quartz and Sejnowski, 1997). Thus, the trend is towards an interactionist, emergent model of the relationship between genes and the environment. Developments like these mean that a simplistic distinction between human evolution and human history is increasingly difficult and unproductive to make. In the emergence of the modern human condition, biological and cultural evolution have become seamlessly integrated (Ingold, 1996).

Cultural evolution, which is the evolution of technological tools and practices, has replaced biological evolution as the principal arena of change on both the time-scale of individual lifetimes and for longer periods of history. Cultural evolution is intimately bound to the communicative and cognitive capacities of human beings. But these are dialectically created, acquired and expressed through human social practices. These have always been collective in the sense that humans have evolved to live in clan or tribe groups rather than as solitary individuals. However, these groups now exist, when they survive at all, within far larger urban collectives. The demographics of this century are striking here. When the present century began, not much over 10 per cent of the human population lived in cities. As it ends this figure is close to 50 per cent and is growing rapidly. The city is now the most common human habitat.

Lewis Mumford's remarkably far-seeing works, especially *Technics and Civilisation* and *The Culture of Cities*, explore the role of urbanization and technology in cultural evolution (Mumford, 1940, 1968). What he called the eotechnic, paleotechnic and neotechnic phases cover, respectively, tool-making, power generation and information-handling. These phases do not just chart the development of techniques. They track the incorporation of technology into the human condition. Mumford makes two principal observations about this incorporation. First, it makes possible radically new conditions of human association. Secondly, technology is not merely the means by which relatively unchanging human needs are met but is the means by which new ones are created. It helps fundamentally new forms of human activity and desire to emerge.

As human beings collected in large urbanized habitats, the pace of cultural evolution accelerated. According to some accounts this was somewhat like a nuclear reactor as it goes critical. The opportunity to associate and exchange, once over a certain limit, releases social and cultural forces of a new character and strength (Teilhard de Chardin, 1965: 261–76). In doing so, in creating what Bourdieu has called the habitus of human societies, human practices are coordinated and our awareness of the world is shaped at all levels (Bourdieu, 1977, 1984, 1991). In this way, as psychologists interested in culturally created meanings have pointed out, human cognitive abilities, and indeed the capacity for consciousness itself, are created (Bruner, 1990; Wertsch et al., 1995). We acquire not only ways of identifying objects and situations but also ways of evaluating and reacting to them. This is the deeper and perhaps more important meaning of Bourdieu's insight into habitus. Cultures are carried by a system made up from practices, beliefs, artefacts and the very physical layout of the built environment. In contemporary industrialized cultures, this system means urban living. The system creates not only opportunities for human action and association but also the sensitivities and values that go with them.

The dynamics of neo-technical culture have primarily to do with the circulation of information and the creation of meaning. The forces that now, autonomously, shape economic and political life demonstrate that evolution has proceeded beyond material and energy flows to semiotic circulation

(Deleuze and Guattari, 1984, 1988; Elias, 1989; Baudrillard, 1993). These theories are matched by recent reworking of evolutionary theory towards a systems view of biological-cultural interaction (Barkow et al., 1992; Barlow, 1991).

Thus, we now have a clearer picture, if a more complex one, of how biology and culture interact to produce the human condition. Culture is technologized and developing fastest in the domain of the circulation of information and signs. This has the direct effects that are felt in the 'information economy' and also the more subtle effect of creating a habitus, a system of sensitivities and values. Human beings now developing within this habitus are bound to be fundamentally shaped by it. The assimilation of technology into the human condition with which Mumford was so concerned may be occurring in a more radical manner than he expected.

In what follows, this biological incorporation of technology should be kept in mind as the backdrop to the primary focus, which is how technology is influencing social relations and human experience within the urban habitat. This influence is characteristically reflexive, in that it is likely in the long run to change how the habitat itself is created.

Digital Technology and Social Relations

Cultural evolution, then, is the evolution of artefacts and the skilled practices that go with them. These artefacts are not neutral environmental furniture but agents of cultural transmission. Individuals are fundamentally shaped by technology which draws out and shapes the intrinsic predisposition towards action and the effort after meaning that animates all organisms. In the human case, this drawing out and shaping is primarily social, beginning with the early interactions between infants and the caring adults around them. As individuals develop, and their capacity for independent action increases, these formative interactions occur with artefacts as well as people. This may require another individual to act as intermediary, model or teacher. It may be, however, that an artefact draws out action in and of itself. It is in this sense that objects as well as individuals may be considered to be social.

To call artefacts 'social' may seem to overextend the term but in fact objects are rapidly becoming social in the conventional sense, as digital technology creates machines with social skills (Pickering, 1997). The encounter with these machines is occurring earlier and earlier in human development. This accelerates the assimilation by which, as Mumford and McLuhan pointed out, outer cultural structure becomes inner cognitive structure. This assimilation involves learning skilled practices through which humans interact socially with each other and with artefacts. Here Rogoff's distinction between three levels of human socio-cultural learning provides a framework within which to address the assimilation of technology, especially during human development (Rogoff, 1995).

Rogoff points out that as well as formalized education, knowledge, and the values that go with it, is transmitted through cultural activities, technological practices and everyday social interaction, particularly that between children and the adult world into which they are learning to fit. Rogoff developed a three-level model of qualitatively distinct but overlapping processes through which individuals assimilate socio-cultural practices: apprenticeship, guided participation and participatory appropriation. Apprenticeship is a collective term for activities such as education, training and instruction of all sorts. It is what happens when learners who know they are learning participate with teachers who know they are teaching to develop specific skills and knowledge. In guided participation, explicit instruction is not involved. It is when individuals learn though taking part with others in collective activities that leave a residue of skills and knowledge. Participatory appropriation is personal, tends to be creative and to individualize knowledge. It is when individuals make for themselves a style and a unique set of practices which are the means to achieve goals they have set themselves.

Until recently these forms of learning were predominantly mediated by interaction with people. Increasingly, however, they now involve artefacts which simulate human social skills. Now, as Mumford reminds us, technological artefacts do not only match a static set of human values and goals, they also create new ones. Thus, as artefacts become social, so new values and goals will appear related to the skilled practices that arise in interaction with them. An object may create practices by itself, much as an implement instructs the user how to wield it. It is in this sense that culturally produced opportunities for action, whether provided by objects, situations or people, may be considered to be social (Costall, 1995). While this broadens the meaning of 'social' considerably, this is what is interesting about artefacts with the capacity for social interaction.

Of course, human action and social interaction have been mediated by artefacts since the emergence of modern human beings. What is significant about digital technology is that it allows artefacts themselves to participate in the interaction. The telephone system is not now merely a means to put human beings in touch with each other. It is a quasi-social collection of agents for answering, asking questions, giving information, holding callers, re-trying numbers that were engaged, informing users of another caller waiting, taking messages, re-routing enquiries and so on. It is rapidly becoming an active participant in social exchange. Such participation is now so far advanced that it is no longer merely a matter of technology but of ethics (Lanier, 1995). The concern is that machines with social skills may elicit from the human beings that interact with them a new type of skilled practice that expresses cybernetic rather than human values. As Mumford showed, technology not only amplifies human capacities, it also creates needs, goals and values. At present, digital technology is amplifying human capacities for social interaction. This may create new needs, skills and values that will be expressed in social relations.

An envelope of social relations mediated by digital technology is developing and moving down the age scale. What used to be an activity of adults at work is rapidly becoming what children do at school and in the home while learning, playing and communicating. Cybernetics has moved from automating mundane rationality to automating mundane sociality. Hybrid intelligent action involving people and computers is increasingly part of social situations. Computer systems help people to communicate, design and decide; in the process they have become socialized. For example, supermarket tills not only display prices and product codes, but also provide help and supervision at a quite naturalistic level. They prompt the operator to ask the customer questions, to obtain signatures, to carry out various phases of the transactions and to watch out for errors. The till has access to a lot of information the operator no longer needs to do their job. For example, if a customer asks the price of an item, the operator is quite unlikely to know it, but will obtain it from the machine. The level of prompting from and intervention by the till varies with the skill of the operator. With the checkout closed, operators can be trained by the till itself, with only occasional interventions by a human supervisor required.

Cars now have voices to remind the driver of things and to give advice. Portable computers now accept written and spoken inputs. These, unlike typing, are attached to individuals, making the machine very much more a personal assistant than a mere tool. Buildings are becoming intelligent, with the advent of security systems that require individuals to have electronic keys. Who is in the building, where they are and where they have been are easy to track. Pagers, electronic diaries, remote conferencing and similar systems are converging into integrated systems attached to particular locations. From bringing busy people together to opening doors, these systems are mediating the social *politesse* of communication and mundane social interaction.

Digital technology of increasing sophistication and naturalness of use is also appearing in the home. There is rapid convergence of domestic computers, TVs, faxes, cable and satellite systems, digital audio broadcasting, telephones and other media. This digitized sociability is descending the age scale. The ease with which children get on with computers is a common and disconcerting experience for adults. Being able to operate and to cooperate with technology is not just to do with knowing how to make the video recorder work. It is about feeling at home with machines that are beginning to use language, recognize individuals, make decisions and offer advice. Being at ease with these human simulacra has as much to do with attitudes as with skills. Such simulacra are appearing in all areas of economic and social life. Soon, children will be growing up with systems that will be an integrated resource for education, communication, recreation and entertainment. Their forerunners can already be seen in the increasingly naturalized operating systems of home computers. These prompt, instruct, ask and autonomously act to provide the user with what they want. As systems converge, more sophisticated agents will help people to use them.

Home computers, in addition to paper manuals, now come with extensive CD-ROM and on-line help resources and training tutorials. These remain on the machine and adjust their interventions as the user becomes more skilled. When a difficulty arises adults are likely to ask 'Where's the manual?' Children are more inclined to say 'Let's ask the computer.'

Children who grow up using such systems may on occasion be unconcerned whether or not they are communicating with a human agent. They are likely to take more rapidly than adults to computer-mediated social interaction. It seems clear here that something important is emerging from the way human beings are growing up with artefacts that participate in human social interaction. As social relations are increasingly mediated by machines the habitus that is thereby created will transmit technologized values and sensitivities. This will influence how human beings communicate with each other and how they think of themselves (Lanier, 1995). As well as being an adjunct to human practices, digital technology creates the habitus of attitudes, tastes and modes of social interaction that has come to be called 'cyberspace'. The more direct effects of digital technology, when compared with construction, transport and energy technology, may not be that extensive or distinctive. However, in the longer run, its effects on social life and human relations may have a profound influence on how urban life is lived and how urban habitats are designed.

Digital Design and Collectivity

The convergence that digital technology makes possible has rapidly emerged into the economy, into the home and into social interactions. This is the fluidity of signs and the compression of time and space characteristic of postmodern political economy (Harvey, 1990; Baudrillard, 1993: ch. 5). Home computers equipment are now sold as 'communication centres' where images, texts, signs, sounds and whatever media carry are interchangeable as never before. This shift to a recombinant culture of assembly and quotation is clear in the advertising industry which now recycles images on a shortening time-scale. For example, Ian Dury's 'Reasons to be cheerful' which came out in the late 1970s had by about 1995 become 'Reasons to eat Alpen'. The transformation of mildly subversive pop into yuppie breakfast cereal illustrates neatly the postmodern cliché that ironic reassembly forges new meanings. This is the same digital fluidity that underlies outworking, downsizing, dispersal and the other elements of post-Fordist industrial practices (Winner, 1996).

Digital communication networks permit such rapid circulation and recombination of data, images, messages and practices, that they represent a qualitative break with what had been possible before. In industry, commerce, education and government, this has changed the culture of organization and this change rapidly spread into the wider community. Electronic tools may, initially, merely have been new ways to do old things. Now they have come to mean more than this as digital intelligence has become incorporated into human practice. In doing so it has changed it fundamentally.

Digital technology may prosaically influence the design of urban living simply by amplifying and enhancing existing design practices. The tools of digital technology are now commonplace items in the professional resources of architects, designers and planners. But these tools are not only for visualization, transformation of spaces, surfaces and volumes. They permit circulation, collectivity and consensus. Buildings, environments and arenas for action can be created in virtual space and then entered in order that their layout and their workings can be investigated. Components and activities from other arenas of practice can be inserted and rearranged in virtual models of what is to be built. Designers and clients may walk around the prototype, much as tourists would walk round a town. Alterations, collectively agreed, might be carried out, evaluated and incorporated. Once things are satisfactory, the updated design can be electronically moved into the next phase of the production process.

The impact of digital technology here will have to do with how designs are distributed and used. It means the rapid transfer and reproduction of any data structure: plans, drawings, texts or virtual prototypes. These will migrate and be worked on collectively far more extensively and rapidly than at present. Ways of displaying, storing and distributing architectural work, initially there for commercial reasons, will become a more mobile resource, with clear implications for the collective inhabitation of the built environment. Already digital cities exist that are electronic *Doppelgängers* for real ones. Virtual fantasy environments have existed for some time of course, but these are proving to be far less significant than the hype surrounding them has implied. Of far greater significance are the virtual communities that are growing up within organizations and real locations (Jones, 1995; Donath, 1997). These are now places where designers, students, clients and the general public may inspect, modify and participate in collective work. This has the potential to transform the role of consultation and consensus in the process of planning, designing and building environments.

Increasingly, the design of products passes through an initial virtual phase that digital technology has made possible. This makes for greater sensitivity in the design process. Rapid prototyping followed by market testing and field trials, both real and virtual, are all possible on a greater scale and with more distinctive inputs. This can be seen in contemporary design practices. In the motor industry for example, the production techniques that were put in place to minimize stockholding have been discovered to permit ordering information to counter-flow 'up' the production system to customize cars as they flow 'down'. Similar applications of digital technology are also changing the relationship between architects, engineers and clients (Pickering, 1996). There is no reason why the design of urban habitats could not have the same participatory character. Given that these are shared spaces, there is every reason for this to happen.

Thus, at a more profound level, digital technology may influence urban living by altering social relationships. Urban spaces are social spaces and the significance of cities is the opportunities they create for human meeting and interaction. As social relations become increasingly mediated by technology, the expectations of citizens will change towards the virtual and the distributed. The growth of social venues that provide Internet activities is an illustrative case. These strike a new balance between the local and the global. Real local presence at a cybercafé is required to enter the global world of digital connectivity and interchange. The activities of corporations such as banks have undergone similar changes. The relocations of the early 1990s demonstrated that actual location was still highly significant. The banking areas of cities like London and New York retain their attraction because face-to-face meetings remain important. Once there is real change here, then dispersal will become that much more marked and attractive. In small and medium-sized enterprises, this trend is already clear.

So the significance of networked communication lies as much in the social relations it makes possible as in economic and manufacturing practices. This, along with the entry of socialized artefacts into everyday domestic life will be the route by which digital technology affects urban living in the longer term. These effects are unpredictable in detail, but some broad trends can perhaps be seen if the impact of digital technology is placed within a framework of the wider political and cultural role of technology in general.

Who Is Empowered by Digital Technology?

The explosive growth of digital technology is a media cliché. Its significance for the practical tasks that people routinely face in going about their everyday lives is often exaggerated and distorted. Popular accounts of virtual reality for instance stress the outlandish and the fanciful. The short-lived playful experiments with sex games are a case in point. Their high profile was clearly due to the unmissable media opportunity created by the combination of the latest cultural glamour object, digital technology, with the oldest and most reliable circulation builder, sexuality. Far less attention was given to the vastly more significant applications in training, telematics, design and modelling.

Even in more informed and realistic discussions there can be a surprising lack of reflection about the social and cultural realities surrounding digital technology. A good example is a recent doctoral thesis from MIT's MediaLab entitled 'Inhabiting the virtual city: the design of social environments for electronic communities' (Donath, 1997). The thesis describes some experiments in designing communication systems carried out on the Internet and within the highly developed networked systems of the MediaLab community itself. Throughout, the fundamental metaphor for the community created by networked digital technologies is the city. The thesis begins and ends with it and it is central to its key idea: that successful virtual communities need to be designed so that individuals with identities and histories can inhabit them. Thus: 'There is a close relationship between building and inhabiting, for inhabiting the virtual city is a process of adapting it and adding to it, of building the connections between places and people that make the on-line world a vibrant and vital environment' (Donath, 1997: 97).

The work expresses the excitement and creativity of those who are working to make digital networks a natural adjunct to human social life. The thesis concerns human-scale activities like sending postcards, working with other people and having conversations. The acknowledgements identify the inspiration for the work as the lifelong quest of the author's father 'to formulate the basis of a just society'. The thesis appears to be about turning the rhetoric of techno-enthusiasm into realistic and benign social outcomes. Indeed, the first part is about the sociology of urban living: the meaning of community, the idea of identity and affiliation within a locale and so on. However, apart from a footnote here and there (e.g. Donath, 1997: 12) there is virtually no discussion of just how virtual communities will integrate, or not, with the ones we already have, on the local, national and global levels.

Now the thesis explicitly sets out its objectives as being the design of the virtual community itself, and not about integrating it with any larger social structures. However, the virtual community must integrate with traditional patterns of human social life. If it does not it will remain an isolated specialism or cult. This thesis points to, but leaves unanswered, questions such as 'How does participation in the virtual community relate to citizenship as it conventionally exists?' 'Who will develop the skills and practices to participate?' 'What will these skills and practices do for them?' and so on.

Here the key issues have to do with empowerment, inclusion/exclusion and access. The politics and economics of the Internet are about who, if anyone, is to control traffic and access, who will own and profit from networks and what political role they will play. There is an interesting contrast here between the USA and Europe. Following Gore's initiatives, the USA is more inclined to see networks as a state responsibility instead of being left to free market forces as might have been expected, given the history of private development of transport systems in the USA. In Europe the reverse seems to be the case, as the earlier quotation illustrated and as the Bangemann Report indicated (Bangemann, 1994).

The relative unconcern with the cultural surroundings of networks in Judith Donath's thesis mirrors a similar unconcern in *The Road Ahead*, the highly publicized book on the impact of digital technology by the president of Microsoft Corporation, Bill Gates (Gates, 1995). With little sense of wider historical movements, Gates's eyes are fixed on what lies ahead, where, since he is so instrumental in building it, he quite probably sees further than most. However, he pays little attention to what lies behind. The role of digital technology in the culture of late capitalism is not perhaps as simple as building a highway. While this may be unfair to Gates, who was not trying to make that sort of historical assessment in any case, the lack of concern for broader cultural issues is striking. For example, in an interview in a UK newspaper published roughly at the same time as his book, he speculated on the way the Internet might alter the way in which ideas and communities of interest form within society and whether the Internet would add or detract from cultural diversity. He concluded that it was not his responsibility – he was just going to build it.

As network culture gathers pace, such disingenuous pursuit of technology for its own sake needs to be treated with caution. Despite the rosy presentation of the Internet as democratizing, its significance to a large proportion of the world's population is more likely to be disempowerment and isolation. The community of MediaLab is highly privileged with respect to American society as a whole. Likewise the users of the Internet worldwide have access to resources that will remain well beyond the reach of the vast majority of the world's population, perhaps indefinitely.

Now of course, it is not the fault of those who work in networked communications that there is inequality and injustice. But it needs to be considered whether what is being created will help to make things better or worse. For example, at a meeting in June 1995 on 'Society and the future of computing' Langdon Winner suggested that principles needed to be drawn up that would alert those working in the area to the wider impact of their work on people and society. Attempts were made at the conference to draft such principles and these were subsequently published as the 'Durango Declarations', after the location of the meeting (Harvey and Gross, 1996). The tone of these declarations is that of professional undertakings, like the Hippocratic oath, to use technology for good rather than ill. While broad and diffuse, they show very clearly the growing concern over the social and political impact of digital technology.

Such undertakings are very necessary. Looking at digital technology as part of the culture of late capitalism it is far from clear that its role will be particularly benign or beneficial. It participates in the slide towards the simulacrum which Baudrillard depicts as leading away from autonomy and towards totalitarianism. Digital technology is represented as a means towards greater access, to the opening up of the political process and as the means to ensure accountability. But in the experience of most people, it is just as likely to be accomplishing the opposite. As the trustworthiness of screen images declines, so there grows a loss of confidence that democratic machinery can be trusted. There is growing suspicion that the rhetoric of networks conceals a move towards disempowerment. It may stimulate greater consumption and turnover in economies that already disrupt and damage the biosphere. For the nations that experience this damage, it will do very little that is positive. It is all too easy to represent digital technology as progressive even though the longer-term historical process in which it is bound up is unsustainable.

For example, at the Global Environmental Summit in Rio de Janeiro in 1992 the then President of the USA George Bush was asked by representatives from the Third World to put on the agenda the overconsumption of resources by the technologized countries, especially the USA. Their point was that global warming, deforestation, famine and the loss of cultural diversity were not natural disasters but the geopolitical effect of which the urbanized lifestyle of nations such as the USA was the cause. Therefore, it was suggested, the summit should consider these causes as well as various conservation measures to deal with the effects.

Bush's reply was terse and uncompromising: 'The American Way of Life,' he said, 'is not up for negotiation.' Although not long after this Clinton replaced Bush in the White House, the new administration quickly made clear that whatever else may have changed with the change of presidents, the technological domination by the USA of global political and economic life has not. Moreover, digital technology will be of central importance in this effort (Clinton and Gore, 1993).

Thus the representation of digital communication networks as part of global progress and democratization is highly questionable. They are at present far more likely to contribute to widening the gap between the haves and the have-nots than to narrow it. Within the developed nations clear geopolitical effects of digital technology have appeared, dispersion and downsizing of production being examples. As these effects manifest themselves it is becoming easier to place digital technology as something that has arisen in a context and with a history.

Baudrillard's work charts the transition to a political economy of simulation beyond mere imitation, that is, the fabrication of a copy of something real. The significant step is substituting the simulation for the real: 'Simulation is no longer that of a territory, a referential being or a substance. It is the generation by models of a real without origin or reality: a hyperreal' (Baudrillard, 1983: 2). The virtual cities that digital technology is building are perfect illustrations of this. They demonstrate the space–time and compression that dulls our sensitivity to the destructive consequences of giganticism and overconsumption (Harvey, 1990). Digital technology's role in the spread of Western ideology following the collapse of the Soviet system is subtle coercion. The skills and practices that go with the Internet are obviously a form of intellectual and economic colonialism. This, in concert with the media, entertainment and other aspects of screen culture, is accelerating the already rapid loss of cultural diversity.

Digital technology is now being implicated in the definition of contemporary conditions of personal and political. In a recent lecture Alain Touraine defined the contemporary condition of citizenship as having access to the resources that would permit the citizen to assume whatever identity seemed appropriate (Touraine, 1997). Thus, it seems, technology now enters into the creation of identity itself. This is reminiscent of Walter Benjamin's suggestion that technology, especially the technology of reproduction, would bring fundamental changes to what we think of as the unique and authentic source of identity for any cultural object, be it a person or a work of art. His insights suggest a broader framework within which to understand the way technology is assimilated into the human condition.

Benjamin in Cyberspace

Benjamin's unfinished Arcades Project (original German title: *Das Passagen-Werk*) was fundamentally about the production of human consciousness within the city (Buck-Morss, 1989). His perception of cities, especially of the overwhelming displays in the shopping areas of Paris and Berlin, was central in forming his materialist philosophy of cultural history. The experience of urban living recurs as a motif in his writings. In a psychoanalytic reading of his radio talks for children, Jeffrey Mehlman traces the precursors of the Arcades Project and of his essay 'The work of art in the age of mechanical reproduction' (Benjamin, 1979). In both works Benjamin's experience of the Berlin of his childhood and the Paris of his maturity and exile is a central theme (Mehlman, 1993: 18–21, 67–71). The shaping of consciousness within the city, and how this conditioned the experience of art, the natural world and other people, was close to the core of his work.

Cities express literally in concrete form how the massive overproduction unleashed by mechanical technology leads away from authentic being and towards alienated violence. This overproduction is now amplified by the recombinant culture unleashed by digital technology. In 'The work of art in the age of mechanical reproduction' there is much discussion of film as a living laboratory in which to see, in the simulacrum, the detachment of art from traditional aesthetic and cultural values. It is interesting to note, especially in sections 10 and 15, that the means for this detachment is the separability and transportability of the photographic image. This is precisely the fluidity and mobility that digital technology has unleashed (Winner, 1996; Baudrillard, 1983, see especially the section entitled 'The map precedes the territory').

The city is an artefact for collective living that has been the natural home for the human collective for around 10,000 years. Now, a simulacrum of the city is growing in cyberspace. This virtual city is ramifying through the real city and in the process reproducing it. Cyberspace, the space behind the screen, is virtual and real at the same time. Its virtuality is patent. Its reality lies in the way it is used. This use has seen the rapid production of a new habitus, in Bourdieu's terms. Work is now in hand to make cyberspace powerful and natural enough for its sociability to simulate that of traditional agoras (Mitchell, 1995; Donath, 1997). But simulation is now the condition of hyperreality. In the cities of cyberspace it is possible to visit, look, listen, communicate, meet, buy, browse and simply hang out. The citizen of cyberspace can be participant, explorer, worker, consumer or merely *flâneur*. The practices and values of this new habitus will not only reproduce the economic and social relations of the city but, as Mumford pointed out, will also create new ones.

This is part of the historical transition from value to sign, from production to reproduction, from copies to simulacra and from law to code. This transition is part of Benjamin's central insight. He anticipated the effects that technology would have on aesthetics, on the cultural production of value and on human consciousness. What he foresaw was the significance of the simulacrum, the multiple and mobile sign that seems to stand for a fixed and singular reality but which ends up by standing for nothing and for everything.

The virtual city is a simulacrum into which real life is migrating. The temples of commodity capitalism are not only realized in glass and steel, they are forming everywhere and nowhere in the virtually realized arcades of cyberspace. The *Passagen-Werk*, productive because it is incomplete, is a powerful way to understand how the cities of cyberspace are being built and how they will form the consciousness of those who inhabit them.

Technology creates the urbanized cultural envelope within which human beings exist and, perhaps more significantly, develop. The city is now the natural home of most of humankind. Indeed, the boundary between the natural and the artificial is now more problematic and more contested than ever before (Robertson et al., 1996). As technology becomes more organic and social, so the organic is technologized and increasingly brought within the sphere of human social concern and action. In this suitably ironic exchange of identity, postmodern science, particularly biology and cybernetics, demonstrates that human consciousness is an artefact and that artefacts already have vestiges of consciousness (Langton, 1995).

These developments remind us that the human condition is not exclusively a function of what is in the head or even in the genes. It is inextricably entwined in a system of relations that comprise the biological order of the body, the cultural order outside the body and the mutually evolved processes that integrate them into a seamless whole. What Benjamin recognized was that the system reflects the social dynamics of urban life. These dynamics, Baudrillard claims, are being amplified by the mobility and fluidity of signification created through digital technology.

The early and optimistic humanism of Mumford and Teilhard de Chardin depicted the city as a stage in a progression towards productive human association. The bright visions of Donath and of Mitchell present the virtual city as a continuation of this progression by digital means. But this benign view is not a common appraisal. The urban concentration that Harvey presents as central to the postmodern condition of world culture is darker and more dystopian (Harvey, 1990: ch. 12). Likewise, for Deleuze and Guattari, the flows of information, matter and energy set in motion by the technology of late capitalism are destructive economic and geopolitical forces. While they originate in human action they are now, autonomously, proceeding well beyond human control in what Deleuze and Guattari (1988) have termed the 'machinic phylum'. In doing so they are creating the material conditions for the corporate giganticism and squalid technopolis of *Blade Runner*, within which human consciousness may be formed in the coming century.

The urban environment is now being simulated in virtual space. Design techniques based on digital technology are bringing the built and the images of the unbuilt closer together in space and time. The future arrives earlier than it used to, and with it comes the danger that Baudrillard has identified: in a process of virtualization, the simulacra that pour from screens are becoming the reality of social and political life. This is what Benjamin foresaw – the power of the simulacrum and the dangers created when society cannot contain the forces created by technology. These dangers are actually exaggerated in the hype that surrounds digital technology. The condition of simulation helps conceal its participation in the repression depicted by Mumford (1971) in his later and more pessimistic works such as *The Pentagon of Power*.

Benjamin's Arcades Project concerns the urbanization of consciousness (Buck-Morss, 1989). The virtual technocity is bringing into existence modes of urban living that conceal as never before the disparities that technology creates, on both a local and a global scale. The virtual space opened up by computers and networks is often presented as a social space, an electronic agora in which democratic interaction of autonomous individual citizens will be the means to more openness and justice in governance. But this presentation is a representation, a simulacrum that is in danger of becoming reality. At present it seems more likely that the virtual city will come to stand as a monument to the damage that results when culture cannot contain the effects of technology.

References

Bangemann, H. (1994) *Europe and the Global Information Society*. Report to the June 1994 meeting of the European Council. Brussels: European Commission. (At http://www2.echo.lu/eudocs/en/bangemann.html)

Barkow, J., Cosmides, L. and Tooby, J. (eds) (1992) *The Adapted Mind: Evolutionary Psychology and the Generation of Culture*. New York: Oxford University Press.

Barlow, C. (ed.) (1991) *From Gaia to Selfish Genes: Selected Writings in the Life Sciences*. London: MIT Press. See section 3: 'A systems view of life'.

Baudrillard, J. (1983) *Simulations*, trans. P. Foss. New York: Semiotext(e).

Baudrillard, J. (1993) *Symbolic Exchange and Death*. London: Sage.

Benjamin, W. (1979) 'The work of art in the age of mechanical reproduction', in *Illuminations*, trans. H. Zohn. London: Fontana. pp. 219–53.

Bourdieu, P. (1977) *Outline of Theory of Practice*. Cambridge: Cambridge University Press.

Bourdieu, P. (1984) *Homo Academicus*, trans. Peter Collier. Cambridge: Polity Press.

Bourdieu, P. (1991) *The Logic of Practice*. Cambridge: Cambridge University Press.

Bruner, J. (1990) *Acts of Meaning*. London: Harvard University Press.

Buck-Morss, S. (1989) *The Dialectics of Seeing. Walter Benjamin and the Arcades Project*. Cambridge, MA: MIT Press

Clinton, W. and Gore, A. (1993) *Technology for America's Growth, a New Direction to Build Economic Strength*. Available from the webpage of the US Government Printing Office: <http://www.access.gpo.gov/>

Costall, A. (1995) 'Socialising affordances', *Theory and Psychology*, 5: 467–82.

Deleuze, G. and Guattari, F. (1984) *Anti-Oedipus. Capitalism and Schizophrenia*, trans. Robert Hurley, Mark Seem and Helen R. Lane, preface by Michel Foucault. London: Athlone.

Deleuze, G. and Guattari, F. (1988) *A Thousand Plateaus: Capitalism and Schizophrenia*. London: Athlone.

Donath, J. (1997) 'Inhabiting the virtual city: the design of social environments for electronic communities'. Doctoral thesis, Program of Media Arts and Sciences, School of Architecture and Planning, MIT. Available from Ms Donath's webpage at: <http://judith.www.media.mit.edu/Thesis/>.

Edelman, G. (1992) *Bright Air, Brilliant Fire*. New York: Basic Books.

Elias, N. (1989) 'The symbol theory: an introduction', *Theory, Culture & Society*, 6: 169–217.

European Commission (1994) *Europe's Way to the Information Society: An Action Plan*. Final Report to the European Commission, no. 347, 19 July. Brussels: EC. Available from: <http://www2.echo.lu/eudocs/en/com-asc.html>.

Gates, W. (1995) *The Road Ahead*. London: Viking.

Gibson, W. and Sterling, B. (1992) *The Difference Engine*. London: Gollancz.

Goodwin, B. (1994) *How the Leopard Changed its Spots: The Evolution of Complexity*. London: Weidenfeld & Nicolson.

Haraway, D. (1995) 'Cyborgs and symbionts: living together in the new world order', in C. Gray (ed.), *The Cyborg Handbook*. London: Routledge.

Harvey, D. (1990) *The Condition of Postmodernity*. Oxford: Basil Blackwell.

Harvey, F. and Gross, B. (1996) 'The Durango Declarations Forum', *The Information Society*, 12(1): 73–82.

Ingold, T. (1996) 'The history and evolution of bodily skills', *Ecological Psychology*, 8(2): 171–82.

Johnson, M. (ed.) (1993) *Brain Development and Cognition, A Reader*. Oxford: Basil Blackwell.

Jones, S. (ed.) (1995) *Cybersociety, Computer-Mediated Communication and Community*. London: Sage.

Kingdon, J. (1993) *Self-made Man and his Undoing*. London: Simon & Schuster.

Langton, C. (ed.) (1995) *Artificial Life: An Overview*. London: MIT Press.

Lanier, J. (1995) 'Agents of alienation', *Journal of Consciousness Studies*, 2(1): 76–81.

Maturana, H. and Varela, F. (1992) *The Tree of Knowledge: The Biological Roots of Human Understanding*, revised edn. London: Shambala Press.

Mehlman, J. (1993) *Walter Benjamin for Children*. Chicago: University of Chicago Press.

Mitchell, W. (1995) *City of Bits: Space, Place and the Infobahn*. Cambridge, MA: MIT Press.

Mumford, L. (1940) *The Culture of Cities*. London: Secker & Warburg.

Mumford, L. (1968) *The Future of Technics and Civilisation*. London: Freedom Press.

Mumford, L. (1971) *The Pentagon of Power*. London: Secker & Warburg.

Pearson, I. and Cochrane, P. (1995) '200 futures for 2020', *British Telecommunications Engineering*, 13: 312–18.

Pickering, J. (1996) 'Cyberspace and the architecture of power', *Architectural Design*, 117. (April): 12–19. London: Academy Editions.

Pickering, J. (1997) 'Agents and artefacts', *Social Analysis*, 41(1): 45–62.

Quartz, S. and Sejnowski, T. (1997) 'The neural basis of cognitive development: a constructivist manifesto', *Behavioural and Brain Sciences*, 20(4): 537–71.

Robertson, G., Mash, M., Tickner, L., Bird, J., Curtis, B. and Putnam, T. (eds) (1996) *Future Natural: Nature, Science, Culture*. London: Routledge.

Rogoff, B. (1995) 'Observing sociocultural activity on three planes', in J. Wertsch et al. (eds), *Sociocultural Studies of Mind*. Cambridge: Cambridge University Press.

Sterling, B. (1986) *Mirrorshades: The Cyberpunk Anthology*. New York: Arbor House.

Teilhard de Chardin, P. (1965[1970]) *The Phenomenon of Man*, revised edn with a foreword by Sir Julian Huxley. London: Fontana.

Touraine, A. (1997) 'How shall we live together as different and equal?' Inaugural lecture of the Social Theory Centre, Warwick University, January. Lecture later published as *Pourrons-nous vivre ensemble? Egaux et différents*. Paris: Fayard, 1997.

Valsiner, J. (1991) 'The construction of the mental', *Theory & Psychology*, 1(4): 477–94.

Wertsch, J., del Río, P. and Alvarez, A. (eds) (1995) *Sociocultural Studies of Mind*. Cambridge: Cambridge University Press.

Winner, L. (1996) 'Who will be in cyberspace?' *The Information Society*, 12(1): 63–72.

10 New Technologies: Technocities and the Prospects for Democratization

Douglas Kellner

> men [and women] make their own history, but not under circumstances of their own choosing – Karl Marx

> They who control the Microscopick, control the World – Thomas Pynchon

The current explosion of new technologies and furious debates over their substance, trajectory and effects poses two major challenges to critical social theory and a radical democratic politics: first, how to theorize the dramatic changes in every aspect of life that the new technologies are producing; and, secondly, how to utilize the new technologies to promote progressive social change to create a more egalitarian and democratic society than has been the case for the past two centuries marked by rampant industrial/technological development and the seeming victory of market capitalism over its historical opponents.

In this chapter, I first want to suggest some ways to theorize the current technological revolution without falling into either technological or economic determinism, as well as unwarranted optimism or pessimism. My argument is that one needs to theorize the spread of new technologies and series of transformations that we are undergoing in the context of the current stage of capitalist development, as a crucial part of the global restructuring of capitalism, and thus to think together the current development and imbrication of technology and capitalism. Moreover, one needs to see new technologies as embodying a set of artefacts and practices that themselves can be restructured and reconstituted to carry out individual and group projects, thus rejecting the perspectives of economic or technological determinism. In carrying out this hermeneutical process, one needs to avoid the extremes of either exaggerating or downplaying the autonomous role of technology in this process, as if technology were either the demiurge of the contemporary world, or an unimportant epiphenomenon of a much greater force, such as capitalism or human self-development. In addition, one must avoid two extremes which would either denigrate and demonize technology in the mode of technophobia, or celebrate and deify it in the mode of technophilia. Instead, a critical theory of technology attempts to develop a dialectical optic that avoids one-sided approaches in theorizing and evaluating the genesis of the new technologies and their often contradictory and ambiguous effects.

I also want to develop democratic and activist perspectives on the new technologies, suggesting some ways that they might be used for such things as self-valorization and empowerment, democratization and progressive social transformation, in contrast to strengthening the forces of corporate and state domination. Yet I do not want to fall into the utopianism of the boosters of new technologies, nor the pessimism and defeatism of those who merely see new technologies as an instrument of capital and the state. Finally, I will take on the ideology of the global city and virtual community and will argue that both promoters and critics of these concepts are failing to conceptualize adequately the social and cultural effects of the new technologies, which involve both the development of new social and cultural spaces and perhaps a new public sphere, as well as intensifying class divisions and the creation of 'dual cities' marked by growing differences between the haves and the have-nots.

For a Critical Theory of Technology

In studying the exploding array of discourses which characterize the new technologies, I am bemused by the extent to whether they either espouse a technophilic discourse which presents new technologies as our salvation, that will solve all our problems, or embody a technophobic discourse that sees technology as our damnation, demonizing it as the major source of all our problems. It appears that similarly one-sided and contrasting discourses greeted the introduction of other new technologies this century, often hysterically. To some extent, this was the case historically with film, radio, TV, and now computers. Film, for instance, was celebrated by early theorists as providing new documentary depiction of reality, even redemption of reality, a new art form, new modes of mass education and entertainment – as well as demonized for promoting sexual promiscuity, juvenile delinquency and crime, violence, and copious other forms of immorality. Its demonization led in the USA to a Production Code that rigorously regulated the content of Hollywood films from 1934 until the 1950s and 1960s – no open-mouthed kissing was permitted, crime could not pay, drug use or attacks on religion could not be portrayed, and a censorship office rigorously surveyed all films to make sure that no subversive or illicit content emerged (Kellner, 1997a).

Similar extreme hopes and fears were projected on to radio, television, and now computers. It appears that whenever there are new technologies, people project all sorts of fantasies, fears, hopes and dreams on them, and I believe that this is now happening with computers and new multimedia technologies. It is indeed striking that if one looks at the literature on new technologies – and especially computers – it is either highly celebratory and technophilic, or sharply derogatory and technophobic. For technophilia, one can open any issue of *Wired*, or popular magazines like *Newsweek*, one can read Bill Gates's book *The Road Ahead* (1995), or some of the academic boosters of new technologies like Nicholas Negroponte, Sandy Stone or Sherry Turkle. These folks are sometimes referred to as digerati – intellectuals who boost new technologies – and they also include Alvin Toffler, George Gilder, David Gelernter (incidentally, one of the Unabomber's victims), and countless wannabes who write for the media, specialist journals and other publications and who want to get on the digital bandwagon and extract whatever joys and cultural capital it will yield.

Technophilic politicians include Al Gore and Newt Gingrich in the USA and Tony Blair and his New Labour cohort in the UK. These boosters promise more jobs, new economic opportunities, better education, a bountiful harvest of information and entertainment, and new prosperity in a computopia that would make Adam Smith proud. With powerful economic interests behind the new technologies, one expects the technological revolution to be hyped. And obviously there is also academic capital to be gained through promoting new technologies, so it is not surprising that our colleagues too are promoting these technologies, often in an uncritical fashion. What is perhaps more surprising, however, is the extent of wholly negative discourses on computers and new technologies. In the past years, a large number of recent books on computers, the Internet, cyberspace and the like have appeared by a wide range of writers whose discourse is strikingly technophobic.

One strand of this vast technophobic literature now aimed at computers goes back to 1960s and earlier criticism of technology by Theodore Roszak, Charles Reich, Neil Postman, Jerry Mander, and other long-time critics of media culture and technology, who now aim their anti-technology jeremiads at computers. The same arguments these writers have previously used against technology in general, they are now using against computers, so there is a recycling of much that we've heard before – at least an old-timer like myself who has been fighting the theory wars since the 1960s has heard the siren song demonizing technology many times and has even been seduced upon occasion by its melodies. I was indeed myself something of a technophobe in the 1960s. I always hated machines, never fetishized cars, was indifferent to television, and preferred the joys of reading and excitement of sex and radical politics. Yet was won over to technology in the 1970s by video and media technologies and in the 1980s by computer technologies, and am now attempting to overcome technophobia and develop a dialectical position.

Critiques have emerged from the philosophical community, including Albert Borgmann's *Across the Postmodern Divide* (1994) which claims that new technologies are taking us into the sphere of hyperreality, a term he borrows from Baudrillard, and that we are losing touch with our bodies, with nature, with other people and with focal things and practices. Lorenzo

Simpson's book, by contrast, links technology to modernity (1994), providing another technophobic polemic against how technology is alienating and oppressing us. Postmodern theorists Arthur Kroker and Michael Weinstein have written a book called *Data Crash* (1995) – a highly demonizing and technophobic book which suggests that our culture has crashed, imploded, into hyperreality, and that we've lost touch with reality altogether, that we are ruled by a new virtual class, that we have entered a new stage of virtual capitalism, which comes as a great surprise to those still labouring in sweatshops or factories. But perhaps the most famous technophobe is the Unabomber whose manifesto (Douglas and Olshaker, 1996) is a compendium of anti-technological, technophobic discourses, condemning industrial-technological society in its totality, echoing countercultural writers and theorists like Marcuse, Ellul, and other critics of the technological society who condemned its dehumanizing features, its tendencies towards massification, and its robbing individuals of power and freedom.

Other technophobic missives include Clifford Stoll's *Silicon Snake Oil: Second Thoughts on the Information Highway* (1995) which provides a fascinating contrast with Gates's book, attacking everything that Gates affirms: they provide positive–negative mirror images of each other, both of which are highly one-sided and demonstrate the need for dialectical perspectives. Our comrades on the Left are also enrolled in the ranks of the anti-information technology forces, including Kevin Robins and Frank Webster who advocate neo-Luddism (1986 and Chapters 2 and 3 in this volume), failing to see any progressive aspects to the new technologies which they view primarily as capitalist tools, used by capital to ensure its hegemony and alternately to dominate and overpower or seduce the working class into virtual dreams and techno-fetishism. Thus, while Robins and Webster are aware of the magnitude of the restructuring of capital and of the importance of new technologies in this restructuring, they maintain a gloomy pessimism, believing that these technologies are simply tools of capital hegemony and not resistance and democratization, as I argue below.

Against one-sided technophilic or technophobic approaches, I would argue that we need to develop a critical theory of technology in order to sort out positive and negative features, the up-side and down-side, the benefits and the losses in the development and trajectory of the new technologies. It is necessary to counter promises of technological utopia: that computers will solve all our problems, produce jobs for everyone, generate a wealth of information, entertainment and education, connect everyone, and overcome boundaries of gender, race, class – claims that we hear from Bill Gates, Clinton and Gore, Tony Blair and others. But we also need to counter technological dystopia: that computers are our damnation, that they are vehicles of alienation, mere tools of capital, the state and domination.

Both approaches reveal the need for a dialectical theory that plays off extremes against each other to generate a more inclusive position, indicating how technology can be used both as an instrument of domination and emancipation, as a tool of both dominant societal powers and of individuals struggling for democratization and empowerment. What is at stake, therefore, is theorizing how new technologies can be used both to create a more egalitarian society, and to empower individuals and groups who are currently disenfranchised – a task I will undertake in the following sections of this chapter.

One also needs to distinguish between technology as part of a societal system, as a force of production that inscribes a system of relations of production, and technology as a set of specific instruments and practices used by individuals with their own ends and goals in sight. This involves theorizing connections between technology and the economic, political, cultural and social dimensions of contemporary society, and seeing how technology can be used differently by various groups and individuals in specific contexts. In the current mode of social organization, technology plays such a major role, however, that there has been an explosion of theories of technological determinism which make technology the organizing principle of contemporary society, thus occluding the force and power of economic and political dimensions.

Theories of technological determinism often use the discourse of post-industrial, or postmodern, society to describe current developments. This discourse often develops an ideal-type distinction between a previous mode of industrial production characterized by heavy industry, mass production and consumption, bureaucratic organization and social conformity, and the new post-industrial society characterized by 'flexible production', or 'post-Fordism', in which new technologies serve as the demiurge to a new postmodernity (Harvey, 1989). For postmodern theorists such as Baudrillard, technologies of information and social reproduction (e.g. simulation) have permeated every aspect of society, high tech has created a new social environment and we have left reality and the world of modernity behind, as we undergo an implosion of technology and the human and mutate into a new species (see Baudrillard, 1993 and the analyses in Kellner, 1989b, 1994). For other less extravagant theorists of the technological revolution, we are evolving into a new post-industrial technosociety, culture and condition where technology, knowledge and information are the axial or organizing principles (Bell, 1976).

The post-industrial society is sometimes referred to as the 'knowledge society', or 'information society', in which knowledge and information are given more predominant roles (see the survey in Webster, 1995). It is now certain that the knowledge and information sectors are increasingly important domains of our contemporary moment and, as many have noted, the theories of Daniel Bell and other post-industrial theorists are not as ideological and far off the mark as many of us once argued. But in order to avoid the technological determinism and idealism of many forms of this theory, one should theorize the information or knowledge 'revolution' as part and parcel of a new form of technocapitalism marked by a synthesis of the information and entertainment industries and producing a new form of 'infotainment society'. The limitations of earlier theories of the 'knowledge society' or 'post-industrial society', as well as current forms of the 'information society', revolve around the extent to which they exaggerate the role of knowledge and information and advance an idealist vision that excessively privileges the role of knowledge and information in the economy, in politics and society and in everyday life, downplaying the role of capitalist relations of production, corporate ownership and control, and hegemonic configurations of corporate and state power.

Yet while perceiving the continuities between previous forms of industrial society and the new modes of society and culture described by discourses of the 'post', we should also grasp the novelties and discontinuities (Best and Kellner, 1997). Webster (1995: 5, *passim*) wants to draw a line between 'those who endorse the idea of an information society' and 'writers who place emphasis on continuities'. Although he puts me in the camp of those who emphasize continuities (1995: 188), I would argue that we need to see both continuities and discontinuities in the current societal transformation we are undergoing, that we deploy a both–and logic in this case and not an either–or logic. In other words, we need both to theorize the novelties and differences in the current social restructuring, as well as the continuities with the previous mode of societal organization. Such a dialectical optic is, I believe, consistent with the mode of vision of Marx and neo-Marxists such as those in the Frankfurt School.

I also believe that current conceptions of the information society and emphasis on information technology as its demiurge are by now too limited; the new technologies are modes of information and entertainment and it is becoming harder and harder to separate them. Indeed, as I have been suggesting, the new technologies are much more than merely information technology; they are also technologies of diversion, communication and play, encompassing and restructuring both labour and leisure. Previous forms of culture are rapidly being absorbed within the Internet, and the computer is coming to be a major household appliance and source of entertainment, information, play, communication and connection with the outside world. As clues to the enormity of the transformation going on, as indicators of the syntheses of the knowledge and cultural industries in the infotainment society, I would suggest reflections on the massive mergers of the major information and entertainment conglomerates that have taken place in the USA during the past few years which have seen the most extensive concentration and conglomeration of these industries in history, including:

CBS and Westinghouse: $5.5 billion
MCA and Seagrams: $5.6 billion
Time Warner and Turner: $7.5 billion
Disney/Capital Cities/ABC: $19 billion
GE/NBC/Microsoft: megabillions

These mergers bring together corporations involved in TV, film, magazines, newspapers, books, information databases, computers and other media, suggesting a coming implosion of media and computer culture, of entertainment and information in a new infotainment society. There have also been massive mergers in the telecommunications industry (in the USA between Southwest Bell and California Bell and New York and Atlantic Bell, with a merger between AT&T and major regional systems almost occurring), and with MCI negotiating a $37 billion merger with WorldCom, which topped British Telecommunications and GTE offers). The corporate media, communications and information industries are frantically scrambling to provide delivery for the wealth of services that will include increased Internet access, cellular telephones and satellite personal communication devices and video, film and information on demand, as well as Internet shopping and more unsavoury services like pornography and gambling.

Consequently, the mergers of the immense infotainment conglomerates disclose a synergy between new technologies and media, which combine entertainment and information, undermining such a distinction and requiring an expansion of the concept of the knowledge or information society; thus I introduce the concept of the infotainment society to highlight the imbrications of information and entertainment in the new media and technologies of the present. Together, these corporate mergers and the products and services that they are producing constitute a new form of technocapitalism and new infotainment society that it is our challenge to theorize and attempt to shape to more humane and democratic purposes than the accumulation of capital and corporate/state hegemony.

Theorizing Technocapitalism

This synthesis in the technological and information revolution can be interpreted as a constituent feature of a new infotainment society that itself is part and parcel of a global restructuring of capitalism. Few theories of the information revolution and the new technologies contextualize the structuring, implementation, marketing and use of new technologies in the context of such global restructuring. The ideologues of the information society act as if technology were an autonomous force and either neglect to theorize the interconnections of capital and technology, or use the advancements of technology to legitimize market capitalism (i.e. Gates, 1995). Moreover, critical theorists of the momentous changes in contemporary society often fail to theorize the ways in which the restructuring of capital is connected with technological revolution. Offe (1985) and Lash and Urry (1987, 1994), for instance, see important changes in the economy, polity, culture and society, but view this as a disorganization of capitalism, as its unravelling, rather than as reorganization.

While most of the prophets and promoters of the information society tend to be technological determinists, many of the (neo)Marxists who criticize its ideologies and practices are economic determinists. Both economic and technological determinisms, however, often neglect the role of continuing conflict and struggle, the possibilities of intervention and transformation, and the ability of individuals and groups to remake society to serve their own needs and purposes. In all determinist conceptions, technology and society are passive, humans are left out, active and empowering uses of technology are not considered. With Lewis Mumford (1934), however, we should insist that humans take command of their social circumstances and technology, shape their social environment to enhance their life and use technology to empower themselves and democratize society. Technics – specific technologies – are instruments that can be actively deployed by human beings. Although they are shaped by social forces to serve specific ends, they can be reconfigured, reshaped, and deployed against the purposes for which they are designed (Feenberg, 1991). This is close to what autonomous Marxists call self-valorization, as opposed to capital valorization, using the technics of production and communication against capitalist relations of production and values (see Negri, 1989).

But to avoid the romanticism of voluntarism and humanism, we need to be clear concerning the precise economic, social, political, cultural and technological forces that are currently restructuring every aspect of life and develop strategies based on this knowledge. I introduced the term 'technocapitalism' to describe the synthesis of capital and technology in the present organization of society (Kellner, 1989a). Unlike theories of postmodernity (i.e. Baudrillard) which often argue that technology is the new organizing principle of society, and not socio-economic relations, I propose the term technocapitalism to point to both the increasingly important role of technology and continued primacy of capitalist relations of production. In my view, contemporary societies continue to be organized around production and capital accumulation, and capitalist imperatives continue to dominate production, distribution and consumption, as well as other cultural, social and political domains. Workers continue to be exploited by capitalists and capital continues to be the hegemonic force – more so than ever, since the collapse of communism.

The term technocapitalism points to a configuration of capitalist society in which technical and scientific knowledge, automation, computers and high tech play a role in the process of production analogous to the role of human labour power, mechanization of the labour process and machines in an earlier era of capitalism, while also producing new modes of societal organization and forms of culture and everyday life. We are in a parallel situation, I believe, to the Frankfurt School in the 1930s which was forced to theorize the new configurations of economy, polity, society and culture brought about by the transition from market to state monopoly capitalism. This major restructuring of capital produced new forms of social and economic organization, technology and culture with the rise of giant corporations and cartels, a capitalist state to help organize capitalism whether in a Fascist or a state capitalist form, and with culture industries and mass culture serving as new modes of social control, new forms of socialization, and a new configuration of culture and everyday life (Kellner, 1989a). My thesis is that today media culture and new technologies are vitally transforming every aspect of social life in a process that is creating new forms of society, sometimes described using terms such as the postmodern society, the information society, cybersociety or global post-Fordism.

The concept of technocapitalism thus attempts to avoid technological or economic determinism. The restructuring of capital is producing a very specific new social configuration that I have termed 'the infotainment society' in order to point to the mergers of information and media industries and to the significance of new technologies of information, entertainment and social reproduction. In terms of political economy, the new post-industrial form of technocapitalism is characterized by a decline of the state and increased power of the market, accompanied by the growing power of globalized transnational corporations and governmental bodies and the decline of the nation state and its institutions. To borrow and adapt from Max Horkheimer: whoever wants to talk about capitalism, must talk about globalization, and it is impossible to theorize globalization without talking about the restructuring of capitalism (see Cvetkovitch and Kellner, 1996; Kellner, 1998a).

While knowledge, information and education are probably playing a more important role than ever in the organization of contemporary society, this is because capital is restructuring itself through the implementation of new technologies in every sphere of life. The dangers are that corporate control of knowledge, information, entertainment and technology will provide a tremendous concentration of corporate power without any countervailing forces. The ideologues of the technological revolution and information society are forever arguing that education is the key to future prosperity, that education must be made available to all, and that it is thus the top social priority. This would be fine if education were to be expanded and made accessible to more individuals and if it were able to augment the realm of knowledge and literacies, rather than just to serve as a sophisticated enhancement of job training, focusing on transmitting the skills and knowledge that capital needs to expand and multiply.

Yet new technologies are revolutionizing not only labour, production and leisure, but also education and schooling. The past years have seen major implementation of new technologies in the educational process and a fierce debate over how to deploy them, how to make them accessible to everyone, and whether they are enhancing or destroying education. Whether new technologies will ultimately improve or harm education is not yet decidable, but it is clear that individuals need to develop intensified computer literacy, as well as print literary and, I would add, media literacy, social and cultural literacy and ecoliteracy (see Kellner, 1998b). As we approach an increasingly complex new world, we need to expand greatly and rethink education and literacy and to devise strategies to use technology to strengthen and democratize education.

New Technologies and Democratization

Since new technologies are in any case dramatically transforming every sphere of life, the key challenge is how to theorize this great transformation and how to devise strategies to reconstruct technology and society in the interests of democracy, empowerment of individuals, and other positive values. This involves making creative and democratic uses of new technologies, reconstructing technologies to make them serve human needs and purposes better, and implementing technology in a process of democratization and progressive social change. Obviously, radical critiques of dehumanizing, exploitative, and oppressive uses of new technologies in the workplace, educational sphere, public arena and everyday life are more necessary than ever, but so are strategies that use these instruments to rebuild our cities, schools, economy and society. In previous articles (Kellner, 1995, 1997b), I have argued that new technologies are creating a new public sphere, a new realm of cyberdemocracy, and are thus challenging public intellectuals to gain technoliteracy and to make use of the innovative instruments of communication to promote progressive causes and social transformation. I want to focus, therefore, in the remainder of this section on how new technologies can be used for increasing democratization and empowering individuals.

Given the extent to which capital and its logic of commodification have colonized ever more areas of everyday life in recent years, it is astonishing that cyberspace is by and large decommodified for large numbers of people – at least in the overdeveloped countries such as the USA. In the US, government and educational institutions, and some businesses, provide free Internet access and in some cases free computers, or at least workplace access. With flat-rate monthly phone bills (which I know do not exist in much of the world), one can thus have access to a cornucopia of information and entertainment on the Internet for free, one of the few decommodified spaces in the ultra-commodified world of technocapitalism.

Obviously, much of the world does not even have a telephone service, much less computers, and there are vast inequalities in terms of who has access to computers and who participates in the technological revolution and cyberdemocracy today. Critics of new technologies and cyberspace repeat incessantly that it is by and large young, white, middle- or upper-class males who are the dominant players in the cyberspaces of the present, and while this is true, statistics and surveys indicate that many more women, people of colour, seniors and people from varied class sectors are becoming increasingly active.[1] Moreover, it appears that computers are becoming part of the standard household consumer package and will perhaps be as common as television sets by the beginning of the next century, and certainly more important for work, social life and education than the TV set. In addition, there are plans afoot to wire the entire world with satellites that would make the Internet and communication revolution accessible to people who do not now have telephones, televisions or even electricity.

However widespread and common – or not – computers and new technologies become, it is clear that they are essential to labour, politics, education and social life, and that people who want to participate in the public and cultural life of the future will need to have computer access and literacy. Moreover, although there is the threat and real danger that the computerization of society will increase the current inequalities and inequities in the configurations of class, race and gender power, a democratized and computerized public sphere might provide opportunities to overcome these inequities. I will accordingly address below some of the ways that oppressed and disempowered groups are using the new technologies to advance their interests and progressive political agendas. But first I want to dispose of another frequent criticism of the Internet and computer activism.

Critics of the Internet and cyberdemocracy frequently point to the military origins of the technology and its central role in the practices of dominant corporate and state powers. Yet it is amazing that the Internet for large numbers is decommodified and is becoming more decentralized, becoming open to more and more voices and groups. Thus, cyberdemocracy and the Internet should be seen as a site of struggle, as a contested terrain, and oppositional intellectuals and activists should look to its possibilities for resistance and circulation of struggle. Dominant corporate and state powers, as well as conservative and rightist groups, have been making serious use of new technologies to advance their agendas and if progressives want to become players in the political battles of the future they must devise ways to use new technologies to advance the interests of the oppressed and of forces of resistance and struggle.

There are by now copious examples of how the Internet and cyberdemocracy have been used in progressive political struggles. A large number of insurgent intellectuals are already making use of these new technologies and public spheres in their political projects. The peasants and guerrilla armies struggling in Chiapas, Mexico, from the beginning used computer databases, pirate radio and other forms of media to circulate their struggles and ideas. Every manifesto, text and bulletin produced by the Zapatista Army of National Liberation who occupied land in the southern Mexican state of Chiapas in 1994 was immediately circulated through the world via computer networks. In January 1995, when the Mexican government moved against the Zapatistas, computer networks were used to inform and mobilize individuals and groups throughout the world to support the Zapatistas' struggles against repressive state action. There were many demonstrations in support of the rebels throughout the world, prominent journalists, human rights observers and delegations travelled to Chiapas in solidarity and to report on the uprising, and the Mexican and US governments were bombarded with messages arguing for negotiations rather than repression; the Mexican government accordingly backed off and as of this writing in autumn 1997, they have continued to negotiate with them.

Earlier, audiotapes were used to promote the revolution in Iran and to promote alternative information by political movements throughout the world (see Downing, 1984). The Tiananmen Square democracy movement in China and various groups struggling against the remnants of Stalinism in the former communist bloc used computer bulletin boards and networks, as well as other forms of communication, to circulate information on their struggles. Opponents involved in anti-NAFTA struggles made extensive use of the new communication technology (see Brenner, 1994; Fredericks, 1994). Such multinational networking failed to stop NAFTA, but created alliances useful for the struggles of the future. As Witherford (1999) notes:

> The anti-NAFTA coalitions, while mobilizing a depth of opposition entirely unexpected by capital, failed in their immediate objectives. But the transcontinental dialogues which emerged checked – though by no means eliminated – the chauvinist element in North American opposition to free trade. The movement created a powerful pedagogical crucible for cross-sectoral and cross-border organizing. And it opened pathways for future connections, including electronic ones, which were later effectively mobilized by the Zapatista uprising and in continuing initiatives against maquilladora exploitation.

Thus, using new technologies to link information and practice, to circulate struggles, is neither extraneous to political battles nor merely utopian. Even if material gains are not won, often the information circulated or alliances formed can be of use. For example, two British activists were sued by the fast-food chain McDonald's for distributing leaflets denouncing the corporation's low wages, advertising practices, involvement in deforestation, harvesting of animals and promotion of junk food and an unhealthy diet. The activists counterattacked, organized a McLibel campaign, assembled a McSpotlight website with a tremendous amount of information criticizing the corporation, and found experts to testify and confirm their criticisms. The five-year civil trial, ending ambiguously in July 1997, created unprecedented bad publicity for McDonald's and was circulated throughout the world via Internet websites, mailing lists and discussion groups. The McLibel group claims that their website was accessed over 12 million times and the *Guardian* reported that the site 'claimed to be the most comprehensive source of information on a multinational corporation ever assembled'; it was indeed one of the more successful anti-corporate campaigns (22 February 1996; visit http://www.envirolink.org/mcspotlight/home.html).

Many labour organizations are also beginning to make use of the new technologies. Mike Cooley (1987) has written of how computer systems can reskill rather than deskill workers, while Shoshana Zuboff (1988) has discussed the ways in which high tech can be used to 'informate' workplaces rather than automate them, expanding workers' knowledge and control over operations rather than reducing and eliminating it. The Clean Clothes Campaign, a movement started by Dutch women in 1990 in support of Filipino garment workers, has supported strikes throughout the world, exposing exploitative working conditions (see their website at http://www.cleanclothes.org/1/index.html). In 1997, activists involved in Korean workers' strikes and the Merseyside dock strike in England used websites to gain international solidarity (for the latter see http://www.gn.apc.org/lbournet/docks/).

Most labour organizations, such as the North South Dignity of Labour group, note that computer networks are useful for coordinating and distributing information, but cannot replace the print media that is more accessible to more of its members, face-to-face meetings, and traditional forms of political struggle. The trick is to articulate one's communications politics with actual political movements and struggles so that cyber-struggle is an arm of political battle rather than its replacement or substitute. The most efficacious Internet struggles have indeed intersected with real struggles ranging from campaigns to free political prisoners, to boycotts of corporate projects, to actual labour and even revolutionary struggles, as noted above.

Hence, in comparison to capital's globalization from above, cyber-activists have been attempting to carry out globalization from below, developing networks of solidarity and circulating struggle throughout the globe. To the capitalist international of transnational corporate globalization, a Fifth International of computer-mediated activism is emerging, to use Waterman's phrase (1992), that is qualitatively different from the party-based socialist and communist internationals. Such networking links labour, feminist, ecological, peace and other progressive groups, providing the basis for a new politics of alliance and solidarity to overcome the limitations of postmodern identity politics (on the latter, see Best and Kellner, 1991, 1997, forthcoming).

A series of struggles around gender and race is also mediated by new communications technologies. After the 1991 Clarence Thomas hearings in the USA on his fitness to be Supreme Court Justice, Thomas's assault on claims of sexual harassment by Anita Hill and others (and the failure of the almost all-male US Senate to disqualify the obviously unqualified Thomas) prompted women to use computer and other technologies to attack male privilege in the political system in the USA and to rally women to support female candidates. The result in 1992 was the election of more women candidates than in any previous election and a general rejection of conservative rule.

Many feminists have now established websites, mailing lists and other forms of cyber-communication to circulate their struggles. Likewise, African-American insurgent intellectuals have made use of broadcast and computer technologies to advance their struggles. John Fiske (1994) has described some African-American radio projects in the 'technostruggles' of the present age and the central role of the media in recent struggles around race and gender. African-American 'knowledge warriors' are using radio, computer networks and other media to circulate their ideas and counter-knowledge on a variety of issues, contesting the mainstream and offering alternative views and politics. Likewise, activists in communities of colour –

like Oakland, Harlem and Los Angeles – are setting up community computer and media centres to teach the skills necessary to survive the onslaught of the mediatization of culture and computerization of society to people in their communities.

Obviously, right-wing and reactionary groups can use and have used the Internet to promote their political agendas. In a short time, one can easily access an exotic witch's brew of ultra-right websites maintained by the Ku Klux Klan and a myriad neo-Nazi groups including Aryan Nations and various Patriot militia groups. Internet discussion lists also promote these views and the ultra-right is extremely active on many computer forums, as well as on their radio programmes and stations, public access television programmes and in fax campaigns, video and even rock music production. These groups are hardly harmless, having promoted terrorism of various sorts ranging from church burnings to the bombing of public buildings. Adopting quasi-Leninist discourse and tactics for ultra-right causes, these groups have been successful in recruiting working-class members devastated by the developments of global capitalism which have resulted in widespread unemployment in traditional forms of industrial, agricultural and unskilled labour.

The Internet is thus a contested terrain, used by Left, Right and Centre to promote their own agendas and interests. The political battles of the future may well be fought in the streets, factories, parliaments and other sites of past struggle, but all political struggle is already mediated by media, computer and information technologies and will increasingly be so in the future. Those interested in the politics and culture of the future should therefore be clear on the important role of the new public spheres and intervene accordingly.

Technocities and Everyday Life: Some Inconclusive Conclusions

I hesitate to speculate on the development of global cities and forms of urbanization in the emerging era of technocapitalism. While there is a vast literature on global cities that sees cities replacing the nation state, becoming a vitalized source of business, entertainment, and sometimes community (see Castells 1989), there are other discourses that downplay the importance of the city and offer a more decentralized vision of the coming computopia. A variety of writers including Naisbitt, the Tofflers, and other apologists for information capitalism suggest that labour will be decentralized in the information economy of the future, that work will be done at home, that it will no longer be necessary to work and live in big cities, and so one can enjoy the utopia of the country or small town and the pleasures of new virtual communities, as the giant urban centres of the industrial age decay.

Another affirmative discourse sees the rejuvenation of cities and urban living in healthy and hearty neo-capitalism, in which capital generated by the information revolution will produce a more liveable and sustainable urban environment, leading to a rebirth and regeneration of great urban centres. A contrasting bucolic and communitarian discourse celebrates the decentralization in the new computerized economy that will make possible work at home and life in small towns, pleasant countryside, and the virtual communities of cyberspace. I want to suggest that both of these discourses are ideological, that both grossly underestimate the crisis of the cities, and that neither offers a realistic vision to transform and make liveable the cities of the present.

To begin, it is not yet clear if the new technocapitalist project of the global information/entertainment society will generate sufficient capital to restructure work, rejuvenate the cities, and provide a sustainable life for the vast amount of its denizens. So far, the technocapitalist restructuring has created tremendous wealth for a privileged class, a stock market orgy that has benefited individuals and institutions able to invest and who pursued intelligent investment strategies, while also generating a vast service industry and part-time temporary work for growing numbers of people, as well as growing unemployment and downward mobility for those who are the victims of technological redundancy and corporate downsizing.

The notion of the 'informational city' (Castells) is quite abstract and Castells has never really fleshed it out in either his 1989 book with that title nor his three-volume magnus opus on 'the information age'. Obviously, a tremendous amount of technology goes into the construction, reproduction, and lived spaces of contemporary cities, especially the big urban conglomerates. Of course, information technology plays a major role in the key industries populating the technocities of the present, and enclaves of these cities exhibit a high degree of technoculture and high-tech environment, such as is found in Bill Gates's famous house (1995). But the metropoles of the present era are best described by the 'dual city' concept in which advanced technoculture and environments exist side-by-side with decaying urban scenes marked by overcrowding and homelessness, substandard housing and infrastructures, declining public services, and growing divisions between rich and poor.[2]

Castells's book *The Informational City* (1989) is not really about an 'informational city', which remains an undeveloped concept and at most a futurist projection of a possible new urban form, but rather focuses on how new technologies are restructuring the organization of labour and the spaces of working and living. At most, then, there are informational subcultures such as Silicon Valley, or informational high-tech work environments. There are, of course, the imaginary technocities of the Internet, with their virtual communities, their websites and information-based architectures, their constructed identities, and their new experiences and pleasures, but these technoscapes are at best a utopia of the imagination and at worst an escape from the sufferings and limitations of everyday life in the (post)modern world. For most of us, they are simply a space where we communicate, do research, and perhaps forge a new and uncertain form of social relations, but are not a habitat where we live and die.

Hence, to some extent, the cybertopias and virtual communities celebrated by boosters of the high tech revolution provide compensation for the growing poverty of everyday life under technocapitalism, as well as the tremendous growth of economic poverty and deprivation. Moreover, the concept of virtual urbanism or virtual communities idealizes the social interactions, individual and group activity, and the politics of cyberspace and its cognates. Kevin Robins (Chapter 2 in this volume) points out that ideologues of technocities and virtual communities offer an idealized vision of order and harmony that provides compensatory escape from the messiness, inequalities and disorder of our contemporary situation, and that this vision of virtual life and architecture is congruent with the utopian dreams of Corbusier and other avatars of modern architecture. But many ideologues of the virtual life and Robins and his fellow traveller critics of cyberlife miss the point that part of the attractiveness, fun and democratic participatory openness of cyberlife is precisely the messiness, the disorder, the scrambling of conventional order and codes, and the rough give and take of argumentation, flaming and passionate discussion of issues of common interest. This is not really a community, let alone a city, but is a new form of public space and democratic participation.

Thus, I am comfortable in affirming the vitality and promise of cyberdemocracy in the new public spheres of cyberspace. This space is much more participatory than the space of television and corporate-controlled media culture, it is much more varied and lively than public and cultural spaces of contemporary culture, as well as the older public institutions like the pub, bar or coffee houses – which are in any case closing or being absorbed by giant corporations. Of course, this largely decommodified public space of cyberculture could itself be colonized by capital and be the source of corporate virtual shopping malls and propaganda, as well as highly exploitative mindless entertainment, gambling and pornography.

Yet, in a curious way, the cyberspace disorder often produces forms of order and structure, and its vitality, diversity and belongingness are analogous in some ways to Jane Jacobs's celebration of urban community and her attack on the utopianism of modern architecture (1961). Cyberspaces resemble Jacobs's vital urban communities with their difference and diversity rather than the more harmonious and ordered aestheticized views of Corbusier and the high modernists. The best sites in cyberspace are not harmonious, well-ordered and structured, or even civil and sophisticated, but are full of life and diversity, excitement and adventure, and useful information spiced with diverting entertainment – just like the best urban communities before they were destroyed, or undermined, by crime, corporate restructuring and flight, and the vicissitudes of technocapitalism.

As Marcuse (1998 [1941]) reminds us, although technology can be part of an apparatus of domination, technics can be used as instruments of emancipation or domination. Technologies are human creations which can empower us and enhance human life, although as Mumford (1934) notes, it is imperative for humans to control technologies, to put them in the service of specific goals, to construct technologies that are humanizing and life-enhancing and not dehumanizing and destructive. Technology is the great human adventure of our time and it has highly contradictory effects rendering one-sided approaches hopelessly inadequate to grasp the complexity of the new technological restructuring of the world. Our challenge is to create new technological spaces and worlds and to use technology to transform everyday life. Such a project, I would maintain, is infinitely more constructive than simply denouncing new technologies, or mindlessly celebrating their current configurations.

Still, I hesitate to speculate on the future construction and effects of the new cyberspace, on what sort of communities will be formed, and to what extent these spaces might function as virtual or technocities. It is clear that this is a rapidly mutating space, that it is a site that more and more people are choosing to inhabit, and that in some ways it is replacing, or supplementing, urban and communal life as we've known and lived it over the past centuries. Yet cities are a great product of civilization, based on face-to-face living, shared public spaces, and full bodily presences. It is unlikely that cyberspaces can replace urban spaces, that they can provide habitats for living that are as nourishing and rewarding – or in some cases as destructive – as the cities and communities that are part and parcel of our life histories.

Changes are certainly happening. We are undergoing a Great Transformation, but we are, I believe, too near the beginning of this adventure to determine its structure, social relations, cultural forms and effects. But a technological revolution is going on, it will have massive effects, and it is of utmost importance to us concerning how we will actually use the new technologies – or whether they and the forces that control them will themselves use us in their projects. It is not only necessary for social theorists to theorize the new technologies and their effects and for activists to devise strategies for using the technology to promote progressive political change; it is up to each individual to determine how they will live the new technologies and cyberspaces, how they will themselves deploy them, and whether they will ultimately be empowering or disempowering.

Notes

1 The latest US figures indicate that 70 million American adults over 16 use the Internet, an increase of more than 18 million people in nine months (*New York Times*, 26 August 1998). The study by Nielsen Media Research and CommerceNet estimates that the largest increases were among blacks, American Indians, young adults, and women over 50 in the nine months up to June 1998. 'The report estimated 5.6 million US blacks use the Internet, an increase of 53 per cent from nine months earlier, and 868,000 American Indians are online, an increase of 70 per cent.'

2 Castells (1989) himself frequently uses the 'dual city' concept, of a city divided between rich and poor, haves and have-nots, which he rightly notes is a theme of classical sociology. Soja suggests concepts such as 'Postmetropolis' and auxiliary categories such as Flexcity, Cosmopolis (a globalized and glocalized world city), exopolis, polarcity, carceral city, and simcity to describe the urban scenes of the present (1996: 21f., passim).

References

Baudrillard, J. (1993) *Symbolic Exchange and Death*. London: Sage.
Bell, D. (1976) *The Coming of Post-Industrial Society*. New York: Basic Books.
Best, S. and Kellner, D. (1991) *Postmodern Theory: Critical Interrogations*. London: Macmillan and New York: Guilford Press.
Best, S. and Kellner, D. (1997) *The Postmodern Turn*. New York: Guilford Press.
Best, S. and Kellner, D. (forthcoming) *The Postmodern Adventure*. New York: Guilford Press.
Brecher, J. and Costello, D. (1994) *Global Village or Global Pillage: Economic Reconstruction from Bottom Up*. Boston, MA: South End Press.
Brenner, J. (1994) 'Internationalist labor communication by computer network: the United States, Mexico and NAFTA', unpublished paper.
Castells, M. (1989) *The Informational City: The Space of Flows*. Oxford: Basil Blackwell.
Castells, M. (1993) *End of Millenium*. Oxford: Basil Blackwell.
Castells, M. (1996) *The Rise of the Network Society*. Oxford: Basil Blackwell.
Castells, M. (1997) *The Power of Identity*. Oxford: Basil Blackwell.
Cleaver, H. (1994) 'The Chiapas uprising', *Studies in Political Economy*, 44: 141–57.
Cooley, M. (1987) *Architect or Bee? The Human Price of Technology*. London: Hogarth.
Cvetkovich, A. and Kellner, D. (eds) (1996) *Articulating the Global and the Local: Globalization and Cultural Studies*. Boulder, CO: Westview.
Douglas, J. and Olshaker, M. (1996) *Unabomber. On the Trial of America's Most-Wanted Serial Killer*. New York: Pocket Books.
Downing, J. (1984) *Radical Media*. Boston, MA: South End Press.
Feenberg, A. (1991) *Critical Theory of Technology*. New York: Oxford University Press.
Feenberg, A. (1995) *Alternative Modernity*. Berkeley: University of California Press.
Fiske, J. (1994) *Media Matters*. Minneapolis, MI: University of Minnesota Press.
Fredericks, H. (1994) 'North American NGO networking against NAFTA: the use of computer communications in cross-border coalition building', XVII International Congress of the Latin American Studies Association.
Gates, W. (1995) *The Road Ahead*. New York: Viking.
Harvey, D. (1989) *The Condition of Postmodernity*. Oxford: Basil Blackwell.
Jacobs, J. (1961) *The Death and Life of Great American Cities*. New York: Random House.
Kellner, D. (1989a) *Critical Theory, Marxism and Modernity*. Cambridge: Polity Press and Baltimore: Johns Hopkins University Press.
Kellner, D. (1989b) *Jean Baudrillard: From Marxism to Postmodernism and Beyond*. Cambridge: Polity Press and Palo Alto, CA: Stanford University Press.
Kellner, D. (1990) *Television and the Crisis of Democracy*. Boulder, CO: Westview Press.
Kellner, D. (1992) *The Persian Gulf TV War*. Boulder, CO: Westview Press.

Kellner, D. (ed) (1994) *Jean Baudrillard: A Critical Reader*. Oxford: Basil Blackwell.
Kellner, D. (1995) 'Intellectuals and new technologies', *Media, Culture, and Society*, 17: 201–17.
Kellner, D. (1997a) 'Hollywood and society: critical perspectives', in J. Hill (ed.), *Oxford Encyclopaedia of Film*. Oxford: Oxford University Press:
Kellner, D. (1997b) 'Intellectuals, the new public spheres, and technopolitics', *New Political Science*, 169–88.
Kellner, D. (1998a) 'Globalization and the postmodern turn', in R. Axtmann (ed.), *Globalization and Europe*. London: Cassell/Pinter.
Kellner, D. (ed.) (1998b) *Herbert Marcuse: War, Technology and Fascism*. London and New York: Routledge.
Lash, S. and Urry, J. (1987) *The End of Organized Capitalism*. Cambridge: Polity Press.
Lash, S. and Urry, J. (1994) *Economies of Signs and Space*. London: Sage.
Marcuse, H. (1998) 'Some social implications of modern technology' in D. Kellner (ed.) *Herbert Marcuse: War, Technology and Facism*. London and New York: Routledge.
Moody, K. (1988) *An Injury to One*. London: Verso.
Mumford, L. (1934) *Technics and Civilization*. New York: Harcourt Brace.
Negri, A. (1989) *The Politics of Subversion: A Manifesto for the Twenty-First Century*. Cambridge: Polity Press.
Offe, C. (1985) *Disorganized Capitalism*. Cambridge: Polity Press.
Soja, E. (1996) *Third Space*. Cambridge, MA and Oxford: Blackwell Publisher.
Stoll, C. (1995) *Silicon Snake Oil: Second Thoughts on the Information Highway*. New York: Double Day.
Waterman, P. (1990) 'Communicating labor internationalism: a review of relevant literature and resources', *Communications: European Journal of Communications*, 15, (1/2): 85–103.
Waterman, P. (1992) *International Labour Communication by Computer: The Fifth International?* Working Paper Series 129. The Hague: Institute of Social Studies.
Webster, F. (1995) *Theories of the Information Society*. London and New York: Routledge.
Webster, F. and Robins, K. (1986) *Information Technology: A Luddite Analysis*. Norwood, NJ: Ablex.
Witherford, N. (1999) *Cyber-Marx: Cycles and Circuits of Struggle in High-Technology Capitalism*. Champaign, IL: University of Illinois Press.
Zapatistas Collective (1994) *Zapatistas: Document of the New Mexican Revolution*. New York: Autonomedia.
Zuboff, S. (1988) *In the Age of the Smart Machine: The Future of Work and Power*. New York: Basic Books.

Afterword: Back to the Future?

John Downey

Perhaps one of the least surprising conclusions that may be drawn from this book is that, despite the fanciful futurology which envelops the debate about information and communication technologies in academia and journalism, almost all of the positions adopted are really quite old. Julian Stallabrass remarks that what presents itself as, and is indeed widely perceived to be, the cutting edge is actually a replay of modernism. The rhetoric of revolution embraced by techno-boosters calls upon a variety of sources. Enlightenment, pre-modern, evolutionist and modernist metaphors are repeatedly mobilized to promote the new. There is a sense in which this historical vocabulary could not possibly do justice to the present and some of the preceding chapters busy themselves with an ideology critique of this discourse warning of the exclusions of the past being repeated in the future. This is extremely valuable. However, perhaps this does not tell us about the new. It does not provide us with a new vocabulary: instead it shows the dangers of using the old one.

An insight offered in many chapters is the importance of the complex relationship between ICTs, globalization and political possibility. A very common argument is that ICTs serve to accelerate the process of globalization. This is a process which strengthens capital *vis-à-vis* both nation states and labour movements. Consequently, the room for manoeuvre of social democratic regimes is seen as strictly limited, as any attempt to redistribute wealth significantly would see a flight of capital and employment. One must consider, however, whether this is an ideology of globalization which serves to restrict the imaginations of would-be reformers. It seems that many social democratic parties have bought entirely the globalization thesis, effectively turning them into neo-liberal parties by default. For the UK government, for instance, education, education, education is in reality deeply subordinate to globalization, globalization, globalization.

Globalization is both a reality and an ideology. Capital is more mobile and ICTs do enhance the command and control functions of metropolitan centres. It is an ideology because the response to it is seen as neo-liberal economic policy together with investment in education to produce the workers of the global economy. If capital is becoming more mobile, so too must national governments and labour movements. However, the mobility of capital may serve to emphasize the differences between national labour movements. When Siemens decides to build a chip factory in Newcastle in the UK rather than in Leipzig in Germany it is hard to imagine too many tears being shed on the Tyne, and vice versa when the decision is made to close the factory. While it is difficult to formulate a political response to globalization that would protect social democratic values, at the very least we should see clearly the directions in which globalization is leading.

Here there appears to be a good deal of consensus amongst the contributors to this book. As the power of the nation state is in relative decline so we see the re-emergence of city states and regions. The work of Manuel Castells on the implications of ICTs and the global economy for cities is deservedly influential. The metropolitan centres would seem to be strengthening their grip while other cities and regions struggle to come to terms with the consequences of globalization. Globalization and the development of information and communication technologies spawns an elite class of symbolic analysts. Inequalities between regions and within cities are exacerbated by these tendencies.

Another theme to emerge is the central importance of making connections between virtual and real life, between electronic spaces and urban places. The flight from the real into the virtual world of unlimited possibilities is exposed for the solipsism which it undoubtedly is. Instead of getting caught up in the discourse of new technologies it is essential that we consider the character of the real cities that are being created by a confluence of processes in advanced capitalism. Robins speaks of new information enclosures and this echoes Mike Davis's work on Los Angeles where he speaks of the creation of a Fortress LA. The symbolic analysts are symbolically and physically separating themselves from the rest of the cities in which they live. Again this is hardly a novelty. The means with which it is being carried out are new.

Many contributors have sought to examine the theoretical underpinnings of policy development in Europe and North America. Webster points out the importance of Daniel Bell's work even though he is widely perceived as having lost the academic argument. Futurists such as Toffler and Negroponte are also influential even though both the theoretical and empirical foundations of their work have been questioned repeatedly. Why is this so? Perhaps industrialists and politicians are attracted to techno-boosterism although for diverging reasons. Perhaps the rhetoric of an economic and social revolution and the incredible confidence of the techno-boosters is easier for politicians to handle. However, perhaps the fault also lies at the door of critical academics who seek to distance themselves from policy-making. Of course, to get involved in policy debates is difficult and time-consuming and rarely ultimately satisfactory. It also often means having to encounter ambivalence in a very direct form. Critical academics have been excluded from policy-making because they say uncomfortable things. Perhaps they have not been active enough in attempting to influence policy decisions. Again the work of Manuel Castells may be used as a model. He is both a critical intellectual, highlighting the implications of ICTs for economic, social and political inequality, and an adviser to the European Commission, attempting to make sure that these arguments are heard in the cacophony generated by techno-boosters. This may be seen as compromising him. Yet in contrast to much critical work (which is in danger of becoming too comfortable) he has made a real difference. While he has been influential at quite a rarefied level, other strategies, policies and action plans are in the process of being developed in many cities and regions in advanced capitalist economies and it is possible to make interventions.

One of the motivations behind this book was to bring together writers with contrasting judgements concerning the economic, social and cultural implications of the widespread dissemination of information and communication technologies in advanced capitalist societies. The internal debate that is evident is useful in the attempt to clarify which positions are actually being adopted. All of the authors here avoid the two most common takes on ICTs: visions of heaven or hell. The scenarios presented are visions of heaven and hell, although the ratio of ingredients varies from chapter to chapter.

This represents an achievement of one of the objectives of this book, which was to attempt to transcend the absurdities sometimes found in discussions about ICTs. The problems largely stem from crude notions of causation. Either the economy determines all or technology determines all. One produces visions of dystopia; the other visions of utopia. Both share a crude theory of history that is unilinear. To transcend these positions is not too difficult. The way has been shown by much of the social theory of the twentieth century. Once we have got beyond these positions, however, it becomes much more difficult because we have to begin to think about the character of agency in relation to both technology and economy. The romantic invocation of agency in Stephen Graham's chapter is questioned by Kevin Robins. Yes, there is agency but what are its limits? What can people do to make a difference? This requires us to go beyond both humanism and anti-humanism and is a much more difficult terrain to traverse than the arguments about determinism mentioned above.

One of the features of both Stallabrass's and Robins's chapters concerned the oppressiveness of the modernist visions of technocities. These are cities built on an oppressive, instrumental reason. These are cities without dirt. These are cities where dirt is not tolerated. The key contention of Robins's chapter is that these visions are a flight from the real. The project of creating real cities in this manner is increasingly seen as impossible: hence an internal migration to a virtual world which is perceived to be masterable. This is a very important argument especially when one bears in mind the growing inequalities that advanced capitalist societies are at present experiencing. Robins powerfully demonstrates the conservatism of this migration to the virtual which imagines itself as revolutionary.

Robins leaves us with a choice. We can choose reason or unreason, order or disorder. For Robins, obviously, to choose disorder is to choose freedom. Again this is an old argument but one that has been raging in contemporary

European philosophy for the past 30 years or so. The work of Deleuze and Guattari is often invoked in these debates with reference to information and communication technologies. It appears that one can only oppose instrumental reason by embracing irrationalism.

Douglas Kellner eschews such an either–or. He draws on the work of the first generation of the Frankfurt School in his attempt to generate a dialectical critical theory of technology. Indeed, Adorno and Horkheimer's *Dialectic of Enlightenment* (1947) may help us to navigate these waters for they used critical reason to criticize the deployment of instrumental reason. In so doing they endeavoured not to throw the baby out with the bathwater, to adopt a phrase from Adorno's *Minima Moralia* (1951). Rather than either–or Kellner argues for a position which can accommodate a way of thinking which is both–and. The implications of information and communication technologies are not unilinear. The deployment of information and communication technologies may serve to increase control, to restrict and inhibit democracy. But they can be and indeed are being used to counter such control, to reinvigorate and invent a critical public sphere. The dialectical way of thinking enables us to chart this war of position and can help us act accordingly. We are both objects and subjects of new discourses and regimes whose condition of possibility is the development of information and communication technologies.

A somewhat neglected area in this book has been the greater role which ICTs are playing in surveillance. This ranges from closed-circuit television operating in cities and even villages to information contained about citizens on smart cards. The integrated character of many information systems significantly enhances the capability of the state, for example, to monitor the actions of citizens. Deleuze argues that this marks a shift in the character of societies. We no longer live in disciplinary societies where institutions are relatively autonomous from one another, and increasingly live in societies of control, societies characterized by integrated systems of control. There is a danger here of overstating the case through setting up contrasting epistemes. It is, however, a very interesting hypothesis which deserves to be taken seriously and investigated empirically. The fundamental concern here is for liberty. The focus on information and communication technologies is just another chapter in the history of theoretical critique and political resistance.

This is an old debate and an old struggle. A struggle which is often presented as having no relevance to the global present. Many of the contributors to this volume point to the deeply compromised positions of social democratic parties in Europe and the USA with reference to information and communication technologies. The Luddite analysis serves to keep the possibility of a just world alive through critique; as Stallabrass concludes: 'it is possible to see how, in a very different society, this non-human technology might be turned towards the most human of ends'. It is, however, not enough to settle for a critique of the old dominant discourses dressed in a new guise, to wistfully rehearse the old arguments. We need as well to examine empirically the dissemination of information and communication technologies, to understand their newness, to understand their possibilities, so that we may better imagine a society where these technologies are turned towards the most human of ends. We need to embrace the technologies which may cause us so much harm.

Index

Americanization: of Canada, 148–9; of Europe, 123–4
Amsterdam, Digital City, 1, 3, 21–2, 90–2, 136
Angell, I., 40, 43
Arcades Project, 181, 183

Bakis, H., 15
Bangemann Challenge, 136
Bangemann Report, 4, 127, 128, 136, 178
Baudrillard, J., 6, 179, 180, 190
Behrman, J. N., 42
Bell, D., 60, 80, 122, 190, 206
Bell, S., 5–6
Benjamin, W., 108, 181–3
Berger, S., 3
biology, 62; and culture, 170–2
Blair, Tony, 41, 73–4, 78
body, in the city, 52
Bologna, as city superhighway, 136–7
Borgmann, Albert, 188
Bourdieu, P., 171
Boyer, C., 49
Boyson, S., 161
Braverman, H., 75
Burnham, Jack, 113–14, 117
Bush, George, 180

Canada, 5, 140–1, 146–50
Canadian Network for the Advancement of Research, Industry and Education, 146–7
capitalism, 7, 68–9; globalization and, 3, 36–8, 40, 63–7, 194; technocapitalism, 193–4, 200
Carnoy, M., 37
Castells, Manuel, 4, 27, 36, 37, 42, 128, 129–33, 200, 206–7
Celebration City, 50
censorship, 143
China, 159, 196–7
Chirac, Jacques, 123, 124
city/cities: evolution and, 170–2; global corporations and, 41–3, 47; morphology of, 161, 163; *see also* global city; virtual city
City Net, 18–19, 21
city superhighways, 127, 136–7
Clean Clothes Campaign, 197

Commission of the European Communities (CEC): *Cohesion and the Information Society*, 131; *Growth, Employment and Competitiveness*, 125–6; *Living and Working in the Information Society*, 130–1
Commission on Social Justice, 41, 73
communications, 5; community and, 44–7; globalization of, 39, 66–7; via Internet, 101–2, 104, 105; *see also* telecommunications
community, 44–7, 48–9, 81–5; *see also* virtual community
community cable networks, 19–20
competition, regulation and, 137, 149–50, 151
computer art, 3–4, 108, 110–11; idealism in, 112–14; Latham and, 112–13
computer culture, 3–4, 108, 111–18
computer literacy, 76, 129
computers, 174–5, 191, 195–6
content of information superhighway, in Canada, 147–8, 150
control, virtual technology and, 50–1, 62, 84
Cooley, M., 197
Corey, K., 16
critical theory of technology, 186–92
cultural evolution, 6, 170–2
cultural sovereignty, 148–9, 151
cyberdemocracy, 195, 196–9, 201
cyberspace, 10, 83, 175, 181–2, 202; participatory, 173, 201; planning innovation and, 16–24; and public sphere, 135
cyborgs, 48, 62

Davis, M., 132–3
Dawson, M., 36
Dear, M., 54–5
Deleuze, G., 182–3, 208
democratization, 179–80, 195–9
deregulation, 124–5, 126, 127, 135, 137, 141, 143
design, 176–7, 183
deskilling, 75
developing world, 5–6, 157–60, 162; city morphology, 161–2, 164
development, value of technology in, 155–6, 158–60

d'Haenens, L., 5
digital cities, 136–7, 176
digital technology, 169; collectivity and, 175–7; empowerment by, 177–81; social impact of, 179–80; and social relations, 172–5
disorder, 2, 35, 51, 201; globalization and, 53–4, 56
Donath, J., 177–8
Downey, J., 4–5
dual cities, 132–3, 137, 200
Durango Declarations, 179
Dutton, W., 23
dystopianism, 13–14, 50, 51, 112–14

economic determinism, 192–3
economic globalization, 36–7, 130, 132; city regions and, 42–3; nation-states and, 40, 70–3, 76, 79, 130
Economist, The, 66, 70, 77
Edelman, G., 170
education for global economy, 40, 41, 72, 73–4, 75–7, 194
electronic spaces, 10, 20, 23, 25, 201, 206
employment: flexible, 75, 68, 129; impact of ITCs on, 128, 131–2, 199; *see also* labour force
empowerment, 178–80, 197–9
enclosures, 54
Etzioni, A., 46, 81
European Union: common information area, 126–7; ICTs and, 121–2, 137; information society in, 125–9, 142, 144–5, 168–9
exclusion, risks of, 125–6

feminism, and new technologies, 94, 95, 98, 198
finance, globalization of, 66, 67, 70
Fiske, J., 198
Flusty, S., 54–5
Forget, C., 148
Foster, J. B., 36
France, 16, 18, 20, 23
Franke, H. W., 108
freenets, 20, 146

Gates, Bill, 38, 178–9, 189
Geddes, Patrick, 163–6
gender, and Internet use, 91, 93, 96, 98
genetic computing, 117–18
Gilder, G., 84
Gingrich, Newt, 78
global city, 35, 43, 54–7, 199
global economy *see* economic globalization
global positioning, 17–19

globalization, 2, 54, 63–4, 158; of capitalism, 3, 36–8, 40, 63–7, 194; corporate ideology of, 34, 36–8, 39; and destabilization, 53, 54; ICTs and, 128, 129–30, 205–6
Gore, Al, 5, 45, 46, 78, 169
Goytisdo, J., 55
Graham, S., 1, 2, 35, 43, 48, 50–1, 56
Grant Lewis, S., 161
Grauso, N., 123
Guattari, F., 182–3, 208

Habermas, J., 134–5
Hanna, N., 161
Harasim, L., 38, 44
Harrison, P., 154
Harvey, D., 15, 182
Hayek, F. von, 84–5
Head, S., 75, 77
High Level Group of Experts (HLEG), 128
Hill, S., 12
Hutton, W., 66

ICTs (information and communication technologies), 4–5, 9, 34–5, 37, 74–7, 80, 125; effects of 128–38, 208; globalization and, 128, 129–30, 205–6
inequalities in information, 124, 125, 127, 195, 196; regional, 4–5, 131, 137
information: corporate control of, 194; development of, 36–7, 41, 42
information city *see* virtual city
information society, 121, 122–5, 128; in Europe, 125–9, 142, 144–5, 168–9; post-industrialism and, 60–1, 80, 190–1
information superhighway, 38, 39, 78, 80, 139, 151; content of, 140–1, 142, 143; debates about, 92–3; national identity and, 123, 148
infrastructure, 15, 67, 78, 140, 142–4, 169; in Canada, 146–7
innovation in urban policy, 17–24, 26–9
Internet, 2, 62, 81, 138, 168, 179; communication via 101–2, 104, 105; community and, 45–6, 47, 82–4; in developing countries, 158, 159, 160, 161, 162; guide to, 95–7; as site of struggle, 196–9; women and, 3, 91, 93, 95–6, 99–105; *see also* World Wide Web
Iran, 196

Jacobs, J., 201
Japan, 65, 142, 143, 146

Kanter, R. M., 42
Kellner, D., 6–7, 208
Kelly, K., 62

knowledge *see* information
knowledge elite, 37, 40, 41–2, 72
Kroker, A., 189

labour force, for global economy, 40–1, 72, 131–2; *see also* employment
labour organizations, 129, 197–8
Lane, G., 155
Lash, S., 192
Latham William, 3–4, 112–13, 114, 117
Le Corbusier, 3, 49–50, 109, 110, 112
learning cycle, 157, 162–7
Lehr, W., 141
liberalization *see* deregulation
Libicki, M. C., 141
Liddle, R., 41, 73
Litherland, S., 155
local agency, 10, 11–16, 27–8, 43, 50–1
local government, and resistance to information society, 133–4, 135, 137–8
Los Angeles, 132–3

McLibel campaign, 197
McLuhan, Marshall, 6
Manchester Host, 20
Mandelbaum, S., 15
Mandelson, P., 41, 73
Mallory, Robert, 113, 116
Marcuse, H., 201
market, globalization of, 64–5
Massey, D., 53
Mattelart, A., 44
media mergers, 139–40, 191–2
MediaLab, 177, 179
Mehlman, J., 181
Metropolitan Area Networks, 18
Mexico, 196
Mitchell, William, 25, 26, 29, 37–8, 48–9, 50
modernism, 49, 108–9; in computer culture, 111–18
Mongin, O., 52, 56
Moss, M., 42
Mouffe, C., 47
Mulgan, G., 19
Mumford, L., 171, 173, 193, 201
Musil, R., 51, 56

nation-state, globalization and, 39, 40, 41–2, 43, 69–73
Negroponte, N., 38, 123–5, 206
networks, 37–8, 84, 142; empowerment and, 178–9; effect on authority of national governments, 39–40, 42
New Labour, 2, 45, 46, 73–4; American influence on, 40–1, 77–9; and community, 82; and information society, 80

Nigeria, 158
Noam, E. M., 141
Nye, J. S., 36

Odone, C., 45–6
Offe, C., 192
Ohmae, K., 39, 42
on-line services, 1, 22–4, 50
order, virtual technologies and, 50–1, 62, 84, 116
Ostwald, M., 48
Owens, W. A., 36

Pakistan, 158–9
Pickering, J., 6
place, politics of, 39–43
political consensus, 77–80
Poster, M., 39, 45
post-industrialism, 60–1, 80, 190–1
postmodernism, 81, 114, 115, 116
private sector, information society and, 126, 127, 135, 169
production, globalization of, 65–6, 71
public access terminals, 92, 105, 136, 138
public sphere, 5, 134–5, 195

real-time city, 55–6
region states, 42–3
regional inequality, 4–5, 131, 137
regulation, 144; and competition, 137, 149–50, 151
Reich, R., 37, 40, 71–3
Rheingold, H., 46, 83
Robins, K., 2, 12, 25, 189, 201, 207–8
Robinson, W., 53
Rogers, E. M., 94
Rogoff, B., 172–3
Roman, J., 52
Rondinelli, D. A., 42

Samoff, J., 161
Schiller, H., 122–3
Schoolnet (Canada), 146
Sennett, R., 52
Shearer, D., 17
Simpson, L., 188–9
simulation, 180, 181, 183
Sirois, C., 148
Slouka, M., 81
social change, 19–22, 44, 47, 68, 128, 195
social cohesion, 19, 48, 128, 130–1
social determinism, 14–15, 24–5, 50
social disadvantage, 1, 23–4, 125–6, 201
social relations, 19, 35, 173–5, 177
society, technology and, 11–12, 14, 166–7, 190, 193, 195

software production, 142, 145
Sorkin, M., 14
Stallabrass, J., 3–4, 205, 208
Stalley, M., 163
standard-making, 141, 143, 144, 151
state, 126–7, 143; and information superhighway in Canada, 147–9, 151; mass media and, 139–40, 143, 151
Stoll, C., 189
surveillance, 208
symbolic analysts, 37, 40, 72–3

technocapitalism, 192–9
technocity, in developing countries, 158, 159, 160, 161, 165–6
technological determinism, 7, 10, 11–13, 25, 26, 79–80, 122, 190, 193
technological utopianism, 2, 10, 12, 50, 51, 61–2, 84, 85, 111, 155
technophilia, 186, 187–8
technophobia, 188–9, 207–8
telecommunications, 12–16, 25, 131, 147 ; deregulation of 124–5, 126, 127, 135, 137, 141, 143; growth in urban strategies of, 17–18
telematics, 10, 15–16, 26–7; experimentation with, 16–24, 27–9
teleports, 17–18
Thatcherism, 68–9
Thomas, Clarence, 198
Toffler, A., 2, 37, 78–9, 122, 206
Touraine, A., 180
trade unions, 68, 129, 132
transnational corporations (TNCs), 64, 65, 67, 69–70; media, 66–7; ownership of, 77
Turkle, S., 94, 95

United States of America, 5, 7, 180: electronic delivery of public services, 23, 24; information society in, 141, 142, 143–4; media mergers, 139–40, 191–2; network regulation in, 20, 178
urban migration, 153–5; and Internet migration, 160–2
urban places, electronic spaces and, 10, 25–9, 206
urban planning, 15–16, 52, 162–3; Geddes and, 163–5; Le Corbusier and, 49–50, 109
urban policy, 10, 15–16; innovations in, 17–24, 27–9; telematics and, 27–9
urban services, electronic delivery of, 22–4
Urry, J., 192

van den Boomen, M., 96
van Zoonen, L., 3
videotex, 20, 93
Vidler, A., 49
Virilio, P., 14, 52, 56
virtual city, 21–2, 34–5, 42, 48–52, 182, 200
virtual community, 19, 47, 82–4, 101–2, 176, 178, 201

Warren, R., 13
Webster, F., 2, 122, 189, 191
Weiner, N., 116
Weinstein, M., 189
Williams, R., 7
Windows, 3, 109–10
Winner, L., 179
Witherford, N., 197
Woherem, E., 155
women, 91, 92–3: absence from information technology, 93–8; on-line, 99–105
Women's Guide to the Internet, 95–7
Woodmark, J., 113
World Wide Web, 21; in Canada, 148; use in marketing cities, 18–19

Zuboff, S., 197